Multi-Threshold CMOS Digital Circuits
Managing Leakage Power

Multi-Threshold CMOS Digital Circuits
Managing Leakage Power

Mohab Anis
University of Waterloo

Mohamed Elmasry
University of Waterloo

SPRINGER SCIENCE+BUSINESS MEDIA, LLC

Library of Congress Cataloging-in-Publication

Title: Multi-Threshold CMOS Digital Circuits *Managing Leakage Power*
Author (s): Mohab Anis and Mohamed Elmasry
ISBN 978-1-4613-5053-8   ISBN 978-1-4615-0391-0 (eBook)
DOI 10.1007/978-1-4615-0391-0

To

Hussein Anis and Soheir El-Nahas

and

Elizabeth, Carmen, Samir, Nadia and Hassan Elmasry

Contents

List of Figures

List of Tables

Preface

Technology scaling has become one of the driving forces behind the tremendous improvement in performance, functionality, and power of digital integrated circuits.

A major thrust in integrated circuit design is to minimize power dissipation, while still maintaining high performance operation. From an energy efficiency point of view, there is the potential to scale supply voltages to reduce power. But in order to maintain performance and a large enough gate overdrive, threshold voltages must be scaled. Initially, by scaling both the supply voltage and the threshold voltage, the increase in subthreshold leakage power is small compared to the quadratic reduction in the dynamic power supply in modern CMOS technologies. With extreme scaling, however, the increase in leakage current will begin to dominate the reduction in switching energies. Multi-threshold voltage CMOS (MTCMOS) technology has emerged as an increasingly popular technique to control the escalating leakage power, without sacrificing high performance.

This book discusses the Multi-threshold voltage CMOS (MTCMOS) technology, where circuit techniques are demonstrated to effectively reduce leakage power, while attaining performance with adequate noise immunity. Furthermore, this book shows how MTCMOS technology can be used to reduce the power dissipation of dynamic logic, and to ensure the correct functionality of current-steering logic styles.

$Multi - Threshold\ CMOS\ Digital\ Circuits$ is written for students of VLSI design as well as practicing circuit designers, system designers, CAD tool developers, and researchers. The reader is assumed to have a basic knowledge of digital circuit design, and device operation. The text covers a broad range of circuit design techniques, but each chapter stands alone so that the reader can choose topics of interest and read the chapters in any order. At the end of each chapter, a set of additional references are provided for further reading.

The book is composed of seven chapters:

- Chapter 1 provides an introduction to the book.

- Chapter 2 discusses the main sources of power dissipation in CMOS circuits, with an emphasis on leakage power, how it scales with technology, and possible solutions to minimize leakage power.

- Chapter 3 presents embedded MTCMOS combinational circuits.

- Chapter 4 examines MTCMOS combinational circuits that employ high-V_{th} sleep devices.

- Chapter 5 describes the design of sequential circuits in MTCMOS technology.

- Chapter 6 considers dynamic circuits in MTCMOS technology.

- Chapter 7 shows how MTCMOS technology can be used to design current-steering circuits such as current mode logic.

Mohab Anis
Mohamed Elmasry
Waterloo, Ontario
Canada

Acknowledgments

The authors acknowledge, first, the blessings of God Almighty in their lives in general, and during their careers over the last three years when the research reported in this book was performed.

The authors would like to thank Dr. Shawki Areibi of the University of Guelph, Guelph, Ontario, Canada and Mohamed Mahmoud of Texas Instruments, Texas, USA, for their help and useful discussions related to this book.

The financial support of NSERC and MICRONET and the editorial help of Carl Harris are greatly appreciated.

About the Authors

Mohab Anis was born in Montreal, Canada on February 19, 1974. He received a B.Sc. degree (with Honors) in Electronics and Communication Engineering from Cairo University, Egypt in 1997, and M.A.Sc. and Ph.D. degrees in Electrical and Computer Engineering from the University of Waterloo, Ontario, Canada in 1999 and 2003, respectively. Currently, he is an Assistant Professor in Electrical and Computer Engineering at the University of Waterloo. His research interests include integrated circuit design and design automation for VLSI systems in the deep sub-micron regime. During his Ph.D. studies, Dr. Anis was with Nortel Networks, Ottawa, Canada in the Fall of 2000, and was a consultant for the Canadian Microelectronics Corp. (CMC) at the University of Waterloo for three years. Dr. Anis has authored/co-authored a book and 25 papers in international journals and conferences, and was awarded the 2002 International Low Power Design Contest Prize . He is a member of IEEE and ACM.

Mohamed Elmasry was born in Cairo, Egypt on December 24, 1943. He received a B.Sc. degree from Cairo University, Cairo, Egypt, and M.A.Sc. and Ph.D. degrees from the University of Ottawa, Ottawa, Ontario, Canada, all in Electrical Engineering in 1965, 1970, and 1974, respectively. He has worked in the area of digital integrated circuits and system design for the last 35 years. He was at Cairo University from 1965 to 1968, and Bell-Northern Research, Ottawa, Canada from 1972 to 1974. He has been with the Department of Electrical and Computer Engineering, University of Waterloo, Waterloo, Ontario, Canada since 1974, where he is a Professor and Founding Director of the VLSI Research Group. From 1986-1991, he held the NSERC/BNR Research Chair in VLSI design at the university. He has served as a consultant to research laboratories in Canada, Japan, and the United States. He has authored and co-authored over 450 papers and 15 books on integrated circuit design and design automation. He has several patents to his credit. He is founding President of Pico Electronics Inc., Waterloo, Canada. Dr. Elmasry has served in many professional organizations in different positions and received many Canadian and International Awards. He is a Founding Member of the Canadian Conference on VLSI, the Canadian Microelectronics Corporation (CMC), the International Conference on Microelectronics (ICM), MICRONET, and CITO. He is a Fellow of the IEEE, Fellow of the Royal Society of Canada, and Fellow of the Canadian Academy of Engineers.

Chapter 1

INTRODUCTION

With the explosive growth in the portable electronics market over the past decade, the emphasis in VLSI and system design is shifting away from high speed to low power. However, devices such as PDAs (Personal Digital Assistants) and notebook computers still require a large amount of data processing/throughput. For multimedia devices such as PDAs, the predominant goal of low-power design is to minimize the total power dissipation of the system, given data throughput requirements [1]. With portable computing, a low-power design requires a system design that strikes a balance between the speed and power constraints.

There is also the economical issue of power reduction, as the packaging of such data intensive chips becomes more expensive. Although technology scaling into the deep submicron regime has enabled millions of devices to be integrated on a single chip, it has done little to provide low-power solutions. Therefore, the cost advantage of high integration may be degraded by the packaging costs.

Another important issue in VLSI design is circuit reliability. As technology advances, performance can be enhanced at a constant supply voltage. However, electric fields in devices can cause these devices to break down which eventually leads to chip malfunction. Therefore, it is evident that the supply voltage must be scaled down, and designs that offer high performance at such low voltages should be investigated.

Finally, there is an environmental concern. Studies have shown that personal computers in the United States consume two billion dollars of electricity every year, which produce as much carbon dioxide as five million cars do [2]. This trend has led to the development of the concept of *green computers*, which consume lower power, and are therefore environment friendly.

Consequently, over the last decade, reducing the dynamic switching power has been the principle focus in many of the proposed low-power circuit techniques. At that time, the off-state leakage power was negligible compared to dynamic power. However, as technology scales into the deep sub-micron regime, the increase in leakage power can no longer be neglected; in fact, leakage power may even outpace that of dynamic power in future technologies.

In light of the importance of reducing power dissipation in VLSI designs, and in particular leakage power, this book offers several approaches to reduce leakage power by utilizing the Multi-Threshold Voltage CMOS (MTCMOS) technology.

This book is organized as follows: In Chapter 2, the key sources of power dissipation in digital CMOS circuits are reviewed. The focus is on leakage power, and how technology scaling has affected it, and what the most popular techniques to control it are.

Chapter 3 presents combinational blocks implemented in the embedded MTC-MOS scheme. Chapter 4 describes MTCMOS combinational circuits employing high-V_{th} sleep devices. Sequential circuits implemented in MTCMOS are introduced in Chapter 5. Chapter 6 explores the effect leakage current has on the functionality of dynamic domino logic styles. The proposed domino design resolves the historical trade-off between performance and noise margins in conventional domino logic styles. In addition, an MTCMOS implementation of the proposed domino design, which achieved substantial leakage power savings is detailed. Chapter 7 presents the MTCMOS technology for a different design goal than reducing leakage power; MTCMOS is used to implement the widely popular MOS Current Mode Logic. The proposed MTCMOS implementation allows the reduction of the supply voltage, and eliminates the usage of level converters, while achieving a guaranteed saturation operation.

References

[1] A. Chandrakasan and R. Brodersen, "Minimizing Power Consumption in Digital CMOS Circuits," *in Proceedings of the IEEE*, pp. 498–523, April 1995.

[2] G. McFarland and M. Flynn, "Limits of scaling MOSFETs," *Technical Report, Stanford University*, 1998.

Chapter 2

LEAKAGE POWER: CHALLENGES AND SOLUTIONS

2.1. Introduction

The steady down-scaling of CMOS device dimensions has been the main stimulus for the growth of microelectronics and the computer industry over the last three decades. Without down-scaling, the recent proliferation of mobile devices such as portable phones and personal computers, and the tremendous performance of workstations could not have been achieved. The down-sizing of MOSFETs, has increased the number of transistors in a chip, improved the functionality of LSIs, enhanced the switching speed of MOSFETs and circuits, and reduced the power dissipation. In 1975, Gordon Moore [1] predicted that the number of transistors that can be integrated on a single die would grow exponentially with time. This has held true for more than 25 years as shown in Figure 2.1.

Over the last 25 years, device minimum feature size has scaled down from 6μm to the present 0.1μm. Figure 2.1 also exhibits that the number of logic transistors per chip has been quadrupling every three to four years, whereas the speed of microprocessors has been more than doubling every three years, increasing from 2 MHz for the Intel 8080 in the mid-1970s to well over 2 GHz for the present leading-edge chips.

The supply voltage V_{dd} must also continue to scale down at the historic rate of 30% per technology generation in order to keep power dissipation, and power delivery costs under control in future high-performance microprocessor designs. In order to maintain this generational speed enhancement, the device threshold voltage V_{th} must also scale down with V_{dd} so that a sufficient gate

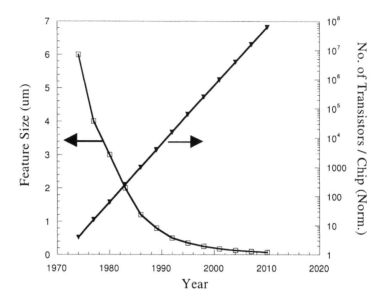

Figure 2.1.   Historical Trend of LSIs

overdrive V_{dd}/V_{th} is maintained. However, this would significantly increase the leakage power.

This chapter highlights the most important power sources in digital CMOS circuits. It then focuses on leakage power as an important power component, and outlines the challenges and possible solutions in the literature.

2.2.   Power Dissipation in CMOS Digital Circuits

In typical CMOS digital circuits, there are two main classes of power dissipation: dynamic power and static power. Dynamic power can be further divided into three components: switching power, short-circuit power, and glitching power; static power has two main elements: DC power and leakage power.

Dynamic Power

The term, *dynamic*, arises from the transient switching behavior of the CMOS circuit. It consists of three components: switching power, short-circuit power, and glitching power. The value of each of these components is a function of the logic style used and the topology of the circuit. Each component will be discussed in the next few sections.

Switching Power

Switching power is defined as the power consumed by the logic gate to charge the output load from the low voltage level "0" to the high voltage output "1". Switching power is expressed as

$$P_{switching} = \alpha F_{operating}.V_{dd}^2.C_L \tag{2.1}$$

where $F_{operating}$ is the operating frequency, V_{dd} is the supply voltage, C_L is the net load capacitance, and α is the switching activity factor of the gate. The net load capacitance C_L consists of the gate capacitance of subsequent gate(s) input(s) connected to the inverter's output, interconnect capacitance, and the diffusion capacitance of the drains of the inverter transistors. Figure 2.2 illustrates the basic capacitive elements of a CMOS inverter.

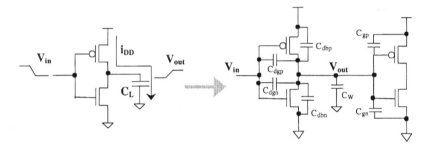

Figure 2.2. Sources of load capacitance in a CMOS inverter

Equation (2.1) indicates that the switching power dissipation increases quadratically with the supply voltage. Thus, reducing the supply voltage is the most effective technique to reduce the power dissipation. This condition is valid only for the logic families with a full output swing. In the case of logic styles employing reduced voltage swings, the switching power is expressed as

$$P_{switching} = \alpha F_{operating}.V_{dd}.V_{swing}.C_L \tag{2.2}$$

where V_{swing} is the logic swing of the digital output.

Short-Circuit Power

In CMOS circuits, short-circuit power is a result of the transient current that flows from V_{dd} to $ground$ when both the NMOS and PMOS devices are turned on during logic transitions. The non-zero rise and fall times of the input signal result in this direct path. The expression quantifying the power consumption in a CMOS inverter due to this condition is given as [2]

$$P_{short-circuit} = \frac{\mu C_{ox}}{12} \frac{W}{L} (V_{dd} - 2V_{th})^3 \tau F_{operating} \qquad (2.3)$$

where μ is the carrier mobility, C_{ox} is the oxide capacitance, τ is the rise/fall time of the input signal, and $F_{operating}$ is the operating frequency. In Equation (2.3), it is assumed that $V_{th_n} = V_{th_p} = V_{th}$, and that the device parameter $= \beta_n = \beta_p = \mu C_{ox} \frac{W}{L}$. Figure 2.3 illustrates the output voltage of a CMOS inverter with short circuit current. For equal input and output edge times, the short-circuit power is minimized and becomes less than 15% of the total dynamic power [3].

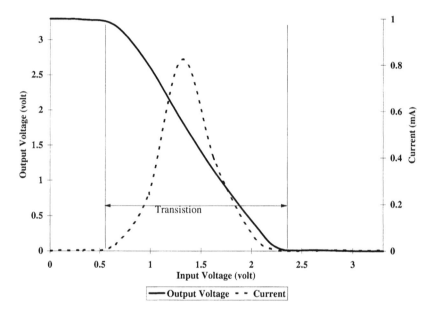

Figure 2.3.   Output of an inverter with short-circuit current

Glitching Power Dissipation

Glitching power is the power dissipated in intermediate transitions during the evaluation of the logic function of the circuit. Unbalanced delay paths are the principle cause of dissipating glitching power. A solution to this problem is buffer insertion to equalize the fast paths in order to achieve a design with balanced delay paths. In addition, a careful layout may reduce the skew among the input signals to each logic gate leading to lower glitching activity.

Static Power

Unlike dynamic power, static power is consumed during the steady-state when no transitions occur. Usually, static power is a small fraction of the total power dissipation. However, as the threshold voltage V_{th} scales down and the number of transistors per chip increases, the static power dissipation becomes more important. There are two major sources of static power dissipation: the DC current and the leakage current of the transistor.

DC Current

A conventional CMOS gate dissipates no DC power, because no path exists between the supply and *ground* in the steady-state. However, some logic styles such as MOS Current Mode Logic (MCML) and Pseudo-NMOS logic consume DC power, which limits their use in low-power designs. Figure 2.4 depicts the pseudo-NMOS logic style. MCML will be discussed in more detail in Chapter 7.

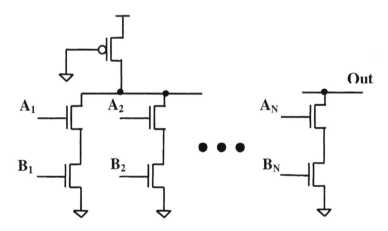

Figure 2.4.   Pseudo-NMOS logic gate

Transistor Leakage Current

There are seven short-channel leakage mechanisms [4] [5], which are denoted in Figure 2.5 [6].

- I_1 is the **reverse bias pn junction leakage**. It has two main components: the minority carrier diffusion/drift near the edge of the depletion region, and the electron-hole pair generation in the depletion region of the reverse bias junction [7].

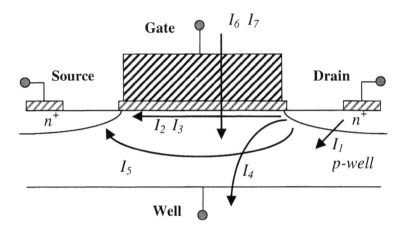

Figure 2.5.   Short-channel transistor leakage current mechanisms: reverse-bias pn junction leakage (I_1), weak inversion (I_2), drain-induced barrier lowering (I_3), gate-induced drain leakage (I_4), punch-through (I_5), gate oxide tunneling (I_6), and hot-carrier injection (I_7) [6]

- I_2 is the **weak inversion current** or **subthreshold conduction current** between the source and drain in a MOS transistor; I_2 occurs when the gate voltage is below V_{th} [7] [8]. The carriers move by diffusion along the surface, and the exponential relation between the driving voltage on the gate and the drain current is a straight line in a semi-log plot [7]. Typically, a weak inversion dominates modern device off-state leakage due to the low V_{th} that is used.

- I_3 represents **Drain-Induced Barrier Lowering** (DIBL). It occurs when the depletion region of the drain interacts with the source near the channel surface to lower the source potential barrier. The source then injects carriers into the channel surface without the gate playing a role.

- I_4 refers to the **Gate-Induced Drain Leakage** (GIDL). GIDL current arises in the high electric field under the gate/drain overlap region, causing a deep depletion [9], and effectively thinning out the depletion width of the drain to the well pn junction. Carriers are generated into the substrate and drain from the direct band-to-band tunneling, trap-assisted tunneling, or a combination of thermal emission and tunneling. The thinner oxide thickness T_{ox} and higher V_{dd} causes a higher potential between the gate and drain which enhances the electric field dependent GIDL.

- I_5 is the **channel punch-through** which occurs when the drain and source depletion regions approach each other and electrically *touch* deep in the channel. Punch-through is a space-charge condition that allows the channel current to exist deep in the sub-gate region, causing the gate to lose control of the sub-gate channel region [7].

- I_6 represents the **oxide leakage tunneling**. The gate oxide tunneling current I_{ox} which is a function of the electric field E_{ox} can cause direct tunneling through the gate [7]. Presently, oxide tunneling current has little effect on the devices in production, but could surpass weak inversion and DIBL as a dominant leakage mechanism in the future as oxides become thinner.

- I_7 is the gate current due to **hot carrier injection**. Short-channel transistors are more susceptible to the injection of hot carriers (holes and electrons) into the oxide [10]. These charges are a reliability risk and are measurable as gate and substrate currents.

Currents I_1 to I_5 are off-state leakage mechanisms, whereas I_6 (oxide tunneling) occurs when the transistor is *on*. However, I_7 can occur in the off-state, but more typically occurs during the transistor bias states in transition. Figure 2.6 summarizes the relative contributions of the main components of intrinsic leakage for a $0.35\mu m$ CMOS technology [11].

2.3.   Impact of Technology Scaling on Leakage Power

As mentioned earlier in this chapter, the supply voltage V_{dd} must also continue to scale down at the historic rate of 30% per technology generation in order to keep power dissipation and power delivery costs under control in future high-performance microprocessor designs. To maintain the speed enhancement for each technology generation, the device threshold voltage V_{th} must also scale down with V_{dd} so that sufficient gate overdrive V_{dd}/V_{th} is maintained. However, reducing V_{th} causes transistor subthreshold leakage current $I_{leakage}$ to increase exponentially. This is demonstrated by Equation (2.4) which formulates the leakage current as [5]

$$I_{leakage} = I_0 e^{(V_{gs}-V_{th}-\gamma V_S+\eta V_{ds})/nV_T}\left(1 - e^{-V_{ds}/V_T}\right) \qquad (2.4)$$

where $I_0 = \mu_0 C_{ox}(W/L)V_T^2 e^{1.8}$, C_{ox} is the gate oxide capacitance, (W/L) is the width to length ratio of the leaking MOS device, μ_0 is the zero bias mobility, V_{gs} is the gate to source voltage, V_T is the thermal voltage which is approximately 26mV at T=300K, n is the subthreshold swing coefficient given by $1+ \frac{C_d}{C_{ox}}$ with C_d being the depletion layer capacitance of the source/drain

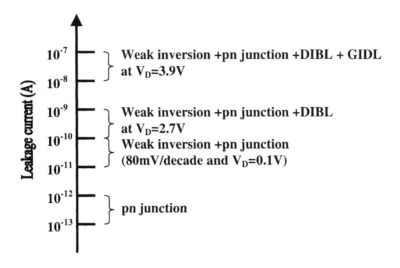

Figure 2.6. Components of $I_{leakage}$ of a $0.35\mu m$ technology for a $20\mu m$ wide transistor (currents from various leakage mechanisms accumulate resulting in a total measured transistor $I_{leakage}$ for a given drain bias [11]).

junctions, γ is the linearized body effect coefficient and η is the DIBL coefficient.

If the body effect and DIBL are neglected, and it is assumed that $V_{ds} \gg V_T$, then Equation (2.4) can be further simplified to the following well known expression [12]:

$$I_{leakage} = \frac{I_0}{W_0} W 10^{\frac{V_{gs}-V_{th}}{S}} \qquad (2.5)$$

where $S=nV_T ln 10$ is the subthreshold slope. For a typical technology with a subthreshold slope S of 100mV/decade, an order of magnitude increase in leakage power occurs with each 100mV decrease in V_{th}.

In practice, there are two key scaling schemes for MOSFET devices. The first scheme is called constant electric field (CE) scaling and was proposed by Dennard et al. in 1974 [13]. In the CE scaling, all the horizontal and vertical dimensions are scaled with the power supply to maintain constant electric fields throughout the device. The second proposed scaling scheme is constant voltage (CV) scaling, proposed by Chatterjee et al. in 1980 [14]. CV scaling maintains a constant power supply, and gradually scales the gate oxide thickness to slow down the growth of fields in the oxide. In industry, the standard

Table 2.1. Impact of Scaling on MOS-Device Characteristics

Parameter	1/S Constant Field Scaling	30% Scaling Field Scaling
Physical Device Dimension	$1/S$	0.7
Supply and threshold Voltage	$1/S$	0.7
$C_{ox} = (\epsilon Area)/T_{ox}$	$1/S$	0.7
Gate Capacitance $= WL/T_{ox}$	$1/S$	0.7
Current $= (W/L)(1/T_{ox})V^2$	$1/S$	0.7
Propagation Delay $= CV/I$	$1/S$	0.7
Frequency	S	1.43
Dynamic Power $= CV^2F$	$1/S^2$	0.5
Leakage Power	exponential	exponential
Energy	$1/S^3$	0.34

scaling methodology has been CE scaling which has resulted in a 30% reduction (1/S=0.7) of all dimensions per generation, where S is the technology scaling factor per generation. Table 2.1 summarizes the technology scaling trends for the CE scaling methodology.

In general, the use of CE scaling, physical device dimensions (width, length, and oxide thickness), as well as voltages (supply V_{dd} and threshold voltages V_{th}), scale by the factor 1/S. Consequently, currents, gate capacitances, and propagation delay also scale by 1/S. Therefore, with the 30% scaling of physical parameters, close to a 50% improvement in frequency has been achieved for each generation. However, this high improvement will gradually decrease when interconnect-dominated delays are taken into consideration as technology advances into the deep submicron regime.

For CE scaling, the resulting switching energy that is dissipated scales by $1/S^3$, where dynamic power scales by $1/S^2$ since the operating frequency increases with scaling. However, for a constant die size, the dynamic power dissipation remains relatively constant with scaling, because the number of switching elements for the same size increases by a factor of S^2. The leakage current increases exponentially with the reduction of V_{th}; furthermore, the total effective width of the devices increases by a factor of S. For example, consider a technology with a V_{th} of 400mV, and a subthreshold slope of 80mV/decade that is to be scaled by 0.7. For a constant die size, scaling provides almost a 43% (S=1.43) improvement in frequency, while increasing the number of devices by a factor of 2. The dynamic power scales by unity, while the leakage power increases by a factor of $1.43 \times 10^{(V_{th}/S(1-0.7))} = 45$ [15].

Although leakage currents are not the dominant component of power dissipation in modern CMOS circuits, it is evident that, as a function of ordinary scaling, the increase in leakage power can soon outpace that of dynamic power in future technologies. Unless circuit techniques or the technology for controlling these leakage currents improve, designers will be forced to scale V_{th} at a lower rate than the power supply in order to prevent further power dissipation. This, though, will tend to reduce the performance gain that could be achieved through scaling over the generations [16].

The active and leakage power trends for Intel's process technologies are represented in Figure 2.7 [17]. The important point is that the leakage power for the 0.25μm technology is 0.1% of the active power, but is approaching 25% of that of the active power in the 0.1μm technology. This further emphasizes that the amount of leakage power could soon outpace that of dynamic power in future technologies.

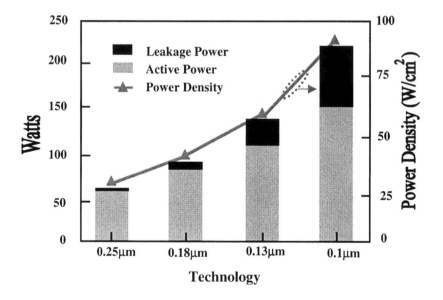

Figure 2.7.   Projected leakage power over generations [17]

2.4. $(V_{dd}\text{-}V_{th})$ **Design Space**

A major goal of the technology scaling trend is to reduce the gate delay by 30%, and to double the transistor density by CE scaling. In addition, a 50% reduction in power can be achieved by a basic shrink at the CE scaling. Power reduction is an important design criterion to satisfy, especially for battery operated systems. As a result, more aggressive $(V_{dd}\text{-}V_{th})$ scaling (which no longer abides by the CE scaling) will become important to minimize power consumption. From an energy efficiency point of view, there is tremendous potential to scale supply voltages to reduce power [18]. Lowering the power supply is the most effective way to reduce power dissipation, because the dynamic switching energy is proportional to the square of the supply voltage as indicated in Equation (2.1). To maintain performance during voltage scaling, the threshold voltage is scaled as well, in order to achieve a large enough gate overdrive. It can be seen how an intrinsic gate speed can be maintained by scaling both V_{dd} and V_{th} as shown in the following :

$$t_{pd} \propto \frac{C_L V_{dd}}{(V_{dd} - V_{th})^\alpha} \tag{2.6}$$

where t_{pd} is the gate propagation delay, and α models short-channel effects [19]. To further illustrate this point, Figure 2.8 displays experimental measured data for a 101 stage ring oscillator consisting of iso-performance curves in the $(V_{dd}\text{-}V_{th})$ space [12]. It is evident that a whole space of $(V_{dd}\text{-}V_{th})$ combinations provides a fixed performance.

With threshold voltage scaling, however, subthreshold leakage currents increase exponentially as quantified in Equation (2.5). Initially, the increase in subthreshold leakage energy is small compared to the quadratic reduction in the dynamic power supply due to V_{dd} scaling for modern CMOS technologies. With further $(V_{dd}\text{-}V_{th})$ scaling, the increase in leakage current can start to dominate the reduction in switching energies. This indicates that there must be an optimum V_{th} point, and consequently, a V_{dd} point for a given delay. Figure 2.9 plots the experimental measurements for the 101 stage ring oscillator, illustrating a minimum energy point as a function of V_{th}, and the corresponding V_{dd} required to maintain performance [12].

For a given process and V_{dd}/V_{th} ratio, the energy efficient V_{th} (and corresponding V_{dd}) point is significantly below the typical threshold levels of today's technologies. This excessive headroom indicates that there is still room for optimal $(V_{dd}\text{-}V_{th})$ scaling to lower overall power dissipation. But, lowering threshold voltages has several undesirable consequences. Noise margins,

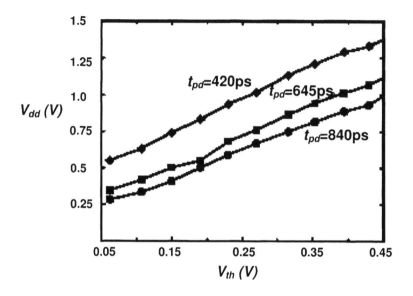

Figure 2.8.   Constant delay curves for V_{dd}-V_{th} span [12]

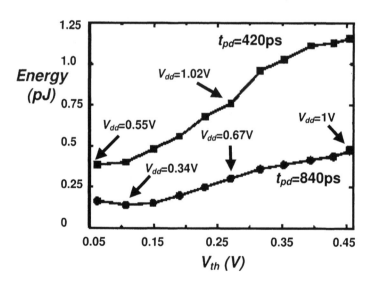

Figure 2.9.   Minimum energy point as a function of V_{dd} and V_{th} [12]

short-channel effects, and V_{th} variations all become worse with lower threshold voltages, and must be carefully balanced with any benefit that may be gained in the overall power dissipation.

2.5. Total Power Management

Although the total power dissipation during the active mode is reduced with scaling, further power savings can be achieved if the subthreshold leakage currents are controlled. As mentioned in the previous section, the scaling theory dictates that subthreshold leakage currents will become a large component of the total power dissipation in future technologies. Similarly, for low-power scaling, the optimum energy point for V_{dd} and V_{th} will correspond to a larger subthreshold component. Moreover, during the standby mode, the standby power dissipation tends to increase since the leakage currents are large as illustrated in Figure 2.10 [15].

Hence, in order to develop a power efficient system, the power dissipation in the active and standby modes must be minimized. To reduce dynamic power in the active period, traditional low-power circuit techniques such as use of pipelining and parallelism, can be employed in order to lower the supply voltage [18]. Circuits implemented with multiple supply voltages [20], and those that minimize effective switching capacitance [21] are good examples of low-power circuit techniques. However, new circuit techniques must be devised to control subthreshold leakage currents in both active and standby modes since these components are becoming more of a problem in modern technologies, and will increase with technology scaling.

Standby subthreshold leakage currents are especially detrimental in *burst* mode type circuits [12], where computation occurs only during short *bursty* intervals, and the system is inactive for the majority of the time, while waiting for the next instruction. This problem is especially severe for portable electronics, where battery power is drained needlessly during long idle periods. As a result, subthreshold leakage reduction techniques during the standby mode can significantly reduce the overall energy consumption for these *burst* mode applications. Leakage power control circuit techniques must therefore be developed.

2.6. Leakage Power Control Circuit Techniques

Over the last decade, many techniques [22] [23] have been reported in the literature to reduce leakage power during the standby mode. Examples of such techniques are: (1) input vector activation, (2) body-biasing, and (3) utilizing the Multi-Threshold CMOS (MTCMOS) technology. In the following sections, these techniques will be briefly examined, and it will be demonstrated the advantage of using the MTCMOS technology over the other two methods, and hence the motivation behind this book.

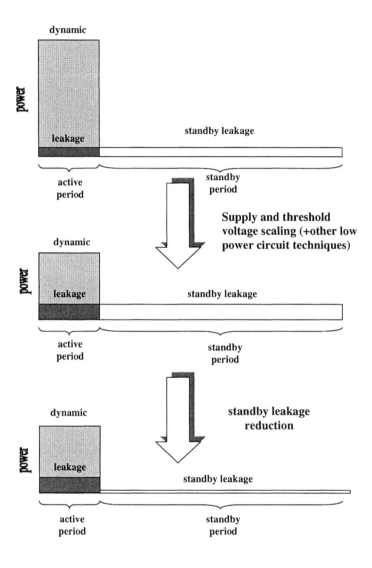

Figure 2.10. Graphical representation showing reduction of total active and standby power [15]

(1) Leakage Control by Input Vector Activation

Leakage control by input vector activation is based on the transistor *stack effect* [24] [25], which refers to the leakage reduction in a transistor stack when more than one transistor is turned *off*. The *stack effect* can be best explained by a two-input NAND gate in Figure 2.11. When both devices Q_1

and Q_2 are turned off, the intermediate node V_M will have a positive voltage due to the small drain current.

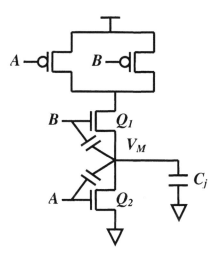

Figure 2.11. Two NMOS stack in a two-input NAND gate

Thus, the gate-to-source voltage of Q_1: V_{gs_1} is negative. The exponential dependence of the leakage current on V_{gs} (Equation (2.5)) greatly reduces the leakage. In addition, the body effect of Q_1, due to $V_M > 0$, further reduces the leakage current as V_{th} increases. In addition, V_M reduces the DIBL, and hence, increases V_{th} of Q_2 which also contributes to the leakage reduction.

As a result, researchers have attempted to make use of the *stack effect* to reduce leakage power by determining the input vector that dissipates the minimum leakage current [26] [27] [28]. Genetic algorithm-based methods and heuristic search methods for leakage bounds were presented in [28] and [29], respectively.

In general, three methods can be used to select the input vector :

- The circuit topology is examined and a *very good* input vector that maximizes the stack effect is determined so that the leakage power is minimized. This method is useful in datapath circuits (such as adders, and multipliers) due to their regular structure.

- Algorithms can be devised to search for the *best* input vector that dissipates the least leakage power. This is especially useful in random logic structures [28].

- A large number of randomly generated input vectors can be applied, and the input vector dissipating the least leakage power is then recorded. In this case, a circuit with n primary inputs has 2^n combinations.

(2) Leakage Control Using Body-Biasing

Researchers have also used body-biasing techniques to control leakage power. The idea behind such techniques can be summarized as follows: In the standby mode, the body V_b of the NMOS device is biased to a voltage lower than *ground* in order to increase the threshold voltage of the NMOS device. Thus, the leakage current is reduced. In the active mode, the body of the NMOS device is biased to *ground* in order to attain the normally low V_{th} value. Thus, the speed of the NMOS device is achieved. Equation (2.7) shows that V_{th} increases when V_b is reduced:

$$V_{th} = V_{th_0} + \gamma(\sqrt{|-2\phi_F + V_{sb}|} - \sqrt{|-2\phi_F|}) \qquad (2.7)$$

where V_{th_0} is the threshold voltage for $V_{sb}=0$, V_{sb} is the source-to-body voltage, ϕ_F is the Fermi potential, and γ is the body-effect coefficient which is a function of the substrate doping, silicon permittivity, and oxide capacitance [30].

Similarly, the PMOS device is biased to a voltage higher than V_{dd} in the standby mode, whereas the body is biased to V_{dd} in the normal active mode. Two examples of body-biasing techniques are the Variable Threshold CMOS (VTCMOS) and the Dynamic Threshold CMOS (DTCMOS).

Variable Threshold CMOS (VTCMOS) : In the VTCMOS technique , the substrate bias voltage is dynamically varied to control the threshold voltage [31] [32]. This technique is depicted in Figure 2.12. All the transistors initially have low V_{th} and the substrate bias is altered for two main reasons: to reduce the leakage current in the standby mode, and to compensate for the V_{th} fluctuations in the active mode, and consequently minimizing the delay variations. VTCMOS has two major drawbacks: (1) the threshold voltage is proportional to the square root of the substrate voltage which requires a large change in the substrate voltage to change V_{th} by effective values, and (2) a charge pump circuit (a substrate bias circuit) is required.

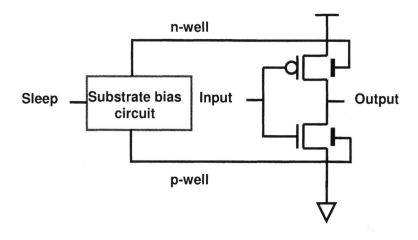

Figure 2.12. Variable Threshold CMOS (VTCMOS)

Other publications have investigated the impact of using the VTCMOS technique on series connected circuits [33], and described the performance and principles of using the VTCMOS technique [34] [35].

Dynamic Threshold CMOS (DTCMOS) : DTCMOS is another kind of body-biasing design technique targeting ultra-low supply voltages [36] [37]. In this scheme, the body and the gate of an MOS transistor are tied together, and the threshold voltage is varied dynamically to suit the circuit's operating state. Figure 2.13 exhibits a DTCMOS inverter.

The threshold voltage of the device is lowered during switching, thereby increasing the transistor drive current. However, the threshold voltage is increased during the standby mode, limiting the leakage current. The DTCMOS technique is limited to supply voltages of less than 0.6V to prevent the forward bias well-to-source junction from conducting large forward bias diode currents. DTCMOS is particularly useful in SOI structures due to the electrical isolation of both the n-well and p-well [38].

(3)Leakage Control by Implementing Multi-V_{th} Designs

An increasingly promising technique to reduce leakage power during the standby modes, while attaining high performance during the active mode, is

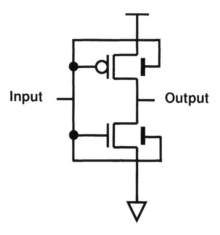

Figure 2.13.   DTCMOS inverter

to utilize multi-V_{th} devices in the circuit implementation [39] [40]. This will be the focus for the remainder of this book. In the following subsections, the implementation of the MTCMOS technology is briefly discussed followed by an overview of some of the multi-V_{th} issues discussed in the book.

Fabrication of MOS devices with Different Threshold Voltages: Multiple threshold voltage devices can be achieved in several ways :

- **Ion Implantation**: To adjust the thresholds by ion implantation, extra masks are required. Presently, this technique is commonly used to modify the threshold voltage [22]. Figure 2.14 plots the threshold voltage at different channel doping densities [41] [42] [43] [44] [45] [46] [47].

  However, the threshold voltage can vary due to the non-uniform distribution of the doping density which makes it difficult to achieve different V_{th}, when the threshold voltages are close to each other.

- **Depositing two different oxide thicknesses** (T_{ox}) : This technique complicates the process, but the large oxide thickness for the high V_{th} can reduce the gate capacitance which is beneficial for the reduction of both the dynamic and subthreshold leakage power [42] [49] [50] [51]. Figure 2.15 shows the threshold voltage for different oxide thicknesses. Moreover, the thicker T_{ox} can suppress gate oxide tunneling, an important concern for ultra-thin T_{ox} technology [7].

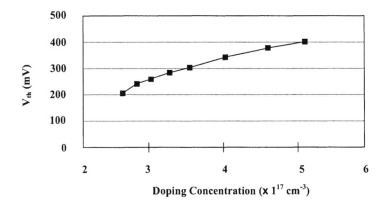

Figure 2.14. V_{th} at different channel doping densities [48]

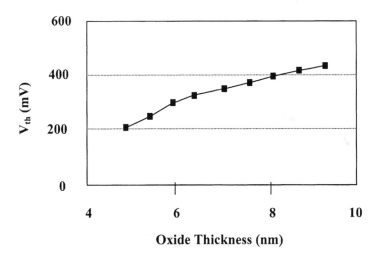

Figure 2.15. V_{th} at different oxide thicknesses [48]

- **Different channel lengths**: There is a V_{th} roll-off due to the Short-Channel Effect (SCE). Therefore, threshold voltages can be achieved by using different channel lengths. Figure 2.16 plots the variation of the threshold voltage with the channel length [52].

  However, the shape of the V_{th} can be very sharp when HALO techniques are used [53]. In such technologies, it is not trivial to control threshold

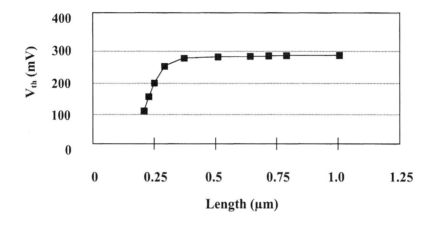

Figure 2.16. V_{th} at different channel lengths [48]

voltages by using multiple channel lengths. Longer transistor lengths for high threshold transistors will increase the gate capacitance which has a negative effect on the performance and power.

- Changing the body or back gate voltage: For bulk silicon devices, the body voltage can be changed to modify the threshold voltage. In this case, the transistors cannot share the same well; Triple well technology is required. It is easier, though, to change the body bias of partially-depleted Silicon-on-Insulator (SOI) devices, since they are isolated naturally.

There are two main ways to utilize the MTCMOS technology to manage leakage power during the standby mode while maintaining performance in the active mode; a dynamic approach and a static approach. The term *dynamic* relates to a design that requires a signal to control certain devices, whereas *static* designs do not employ any control signals.

Leakage Control by Using High-V_{th} Sleep Transistors (Dynamic Approach)

The dynamic approach could be divided further into three kinds which depend on how the sleep transistor is controlled. The control signal could be: a $SLEEP$ signal, a multi-voltage signal, or a boosted $SLEEP$ signal.

- To dynamically control the leakage power dissipated in the CMOS circuit, a control signal ($SLEEP$ signal) is commonly employed to turn off devices in the standby mode ($SLEEP$=0) to minimize the leakage power. This technique is similar to disabling the clock feeding a system during the idle periods in order to cut off the dynamic power dissipation. The MTCMOS circuit [54] [55] depicted in Figure 2.17, employs both low-V_{th} and high-V_{th} transistors. The logic gates are connected to the virtual $ground$ (VV_{ss}) and virtual supply (VV_{dd}); they are implemented by using low-V_{th} devices in order not to degrade the gate's speed, and to operate at low voltages during the active mode, while establishing satisfactory performance. These virtual lines are connected to the main supply while the $ground$ lines are connected through the high-V_{th} transistors.

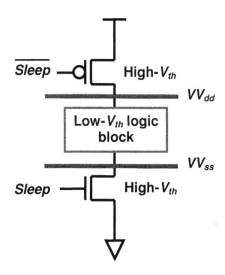

Figure 2.17.   Multi-Threshold CMOS (MTCMOS)

During the standby mode, the sleep signal is disabled to turn off the high-V_{th} devices. This causes the VV_{ss} and VV_{dd} lines to float, limiting the leakage current to that of the high-V_{th} devices. It should be noted that although both PMOS and NMOS sleep transistors are displayed in Figure 2.17, only a single polarity sleep device is actually required to reduce the leakage. Typically, NMOS sleep transistors are more effective, because they have a lower on resistance than that of the PMOS sleep transistor, and subsequently, can be made smaller for the same current drive.

A further modification to the MTCMOS structure has been proposed in [56], where the data is preserved during the standby mode using a conventional static latch implemented with a high-V_{th} called the *balloon* scheme. There are several drawbacks associated with the MTCMOS scheme. First, the multi-threshold voltage technology has to be available for low-V_{th} and high-V_{th} devices. Secondly, the high-V_{th} devices can limit the down-scaling of the supply voltage for ultra low-power applications. Thirdly, the extra latches for storing and the sleep transistors would contribute to extra area overhead. Fourthly, the virtual lines have a much higher impedance than the main supply lines, and will unavoidably bounce which will impact both the delay and noise margins. The delay is influenced by the reduced effective supply voltage and the increased V_{th} due to the body effect. Lastly, the sleep transistors have to be sized large to reduce delay and to minimize the noise bouncing on the virtual lines [57].

- Multi-Voltage CMOS (MVCMOS) [58] [59] shown in Figure 2.18, addresses some of the problems in MTCMOS. MVCMOS does not have high-V_{th} devices to reduce the leakage power. Instead, MVCMOS employs low-V_{th} sleeping transistors whose gate voltages are driven in the sleep mode to larger values than V_{dd} for the PMOS transistor, and less values than $ground$ for the NMOS transistor.

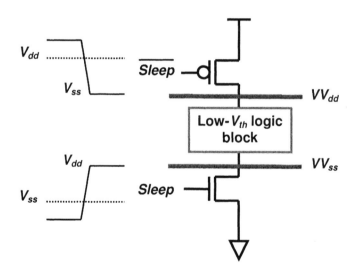

Figure 2.18. Multi-Voltage CMOS (MVCMOS)

This creates a negative value for V_{gs} for both PMOS and NMOS sleep transistors which substantially reduces the leakage current. The elimination of the high-V_{th} transistors allows the supply voltage to be scaled down more than that in the MTCMOS case. Also, since the sleep transistors have a low-V_{th}, then they do not need to be sized up to attain a good performance such as the MTCMOS technique. However, MVCMOS would still be vulnerable to the supply bounce problem, and, in addition, would require positive and negative charge pumps.

- A boosted gate MOS technique also attempts to solve the problems associated with the MTCMOS design [60]. In this scheme, a high-V_{th} device is stacked in series with the original circuit similar to the MTCMOS scheme. However, the high-V_{th} device employs a thick oxide thickness T_{ox} to further reduce leakage, while attaining high speed. To reduce the area overhead incorporated by the high-V_{th} device, the drive current is improved by applying a higher gate voltage than V_{dd} (V_{boost}) in the active mode.

Embedded Multi-V_{th} CMOS Design (Static Approach)

As stated earlier, unlike the dynamic techniques, the static approach does not employ any control signals for power savings. In the static multi-V_{th} technique, gates off the critical paths are designed to operate at a high-V_{th} in order to reduce leakage power without the performance being affected; yet, gates lying on the critical path operate at a low-V_{th} to maintain high performance [61] [62].

Using this design scheme, researchers have proposed several multi-V_{th} designs. For example, in [63], a logic style suggested for DRAMs has high-V_{th} transistors for those devices that are cut off in the standby cycle. In [64], a dual-V_{th} self-timed logic is developed for a possible application to multi-gigabit synchronous DRAMs. Furthermore, high-V_{th} devices have been used in SRAM cells to suppress leakage power, whereas low-V_{th} devices are utilized in the critical paths of memory-cell access [65]. Figure 2.19 illustrates the idea of the embedded multi-V_{th} circuit.

It should be pointed out that in aggressive high-performance low-power circuit topologies that have several balanced critical paths, many gates cannot be slowed down; only a limited leakage reduction can be achieved. Another difficulty associated with the embedded multi-V_{th} technique is that non-critical gates which are assigned to be high-V_{th} can then become critical gates as illustrated in Figure 2.20. This difficulty highlights the need for improved CAD tools to help designers handle the increased design complexity of multi-V_{th} devices.

There are algorithms that optimally assign gates to be either high-V_{th} or low-V_{th} [26] [27] [28] [66] [67] [68]. Typically, they reduce leakage power by 40% to 80% with minimal penalties in critical-path delay compared to low-V_{th}-only implementations. Other techniques control leakage power by simultaneously optimizing for multi-V_{th}, as well as device sizing [62] [69] [70] [71].

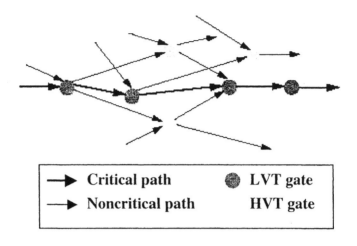

Figure 2.19.   The Concept of Embedded MTCMOS circuits

In addition, leakage power can be reduced by simultaneously optimizing for a multi-V_{th}, as well as a multi-V_{dd} [72] [73].

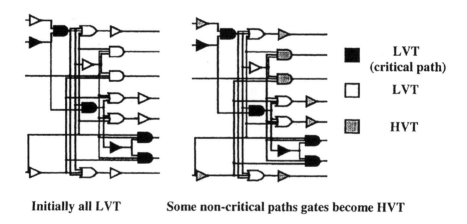

Figure 2.20.   Multi-V_{th} gate partitioning showing how critical paths can change [15]

2.7. Chapter Summary

This chapter starts off with characterizing leakage current, and illustrates how it is impacted by technology scaling. Some leakage power control techniques are then addressed, including leakage control by input vector activation, leakage control using body biasing, and leakage control by implementing multiple V_{th} designs.

References

[1] G. Moore, "Progress in Digital Integrated Circuits," *in Proceedings of the International Electron Devices Meeting*, pp. 11–13, 1975.

[2] A. Bellaouar and M. Elmasry, *Low-Power Digital VLSI Design Circuits and Systems*, Kluwer Academics Publications, 1995.

[3] H. Veendrick, "Short-Circuit Dissipation of Static CMOS Circuitry and Its Impact on the Design of Buffer Circuits," *IEEE Journal of Solid-State Circuits*, vol. 19, no. 4, pp. 468–473, August 1984.

[4] A. Keshavarzi, K. Roy, and C. Hawkins, "Intrinsic Leakage in Low Power Deep Submicron CMOS ICs," *in Proceedings of the IEEE International Test Conference*, pp. 146–155, 1997.

[5] Z. Chen, L. Wei, A. Keshavarzi, and K. Roy, "I_{DDQ} Testing for Deep-Submicron ICs: Challenges and Solutions," *IEEE Design & Test of Computers*, vol. 19, no. 2, pp. 24–33, March-April 2002.

[6] J. Soden, C. Hawkins, and A. Miller, "Identifying Defects in Deep-submicron CMOS ICs," *IEEE Spectrum*, , no. 9, pp. 66–71, September 1996.

[7] R. Pierret, *Semiconductor Device Fundamentals*, Reading,MA: Addison-Wesley, 1996.

[8] Y. Tsividis, *Operation and Modeling of the MOS Transistor*, New York: McGraw-Hill, 1987.

[9] J. Brews, *High Speed Semiconductor Devices*, New York: Wiley, 1990.

[10] T. Hayashi, "Hot Carrier Injection in PMOSFETs," *OKI Technical Review*, pp. 59–62, 1991.

[11] A. Keshavarzi, K. Roy, and C. Hawkins, "Intrinsic Leakage in Deep Submicron CMOS ICs - Measurement-Based Test Solutions," *IEEE Transactions on Very Large Scale Integration (VLSI) Systems*, vol. 8, no. 6, pp. 717–723, 2000.

[12] A. Chandrakasan, I. Yang, C. Vieri, and D. Antoniadis, "Design Considerations and Tools for Low-Voltage Digital System Design," *in Proceedings of the 33rd Design Automation Conference*, pp. 113–118, June 1996.

[13] R. Dennard, F. Gaensslen, H. Yu, V. Rideout, E. Bassous, and A. LeBlanc, "Design of Ion-Implanted MOSFETs With Very Small Dimensions," *IEEE Journal of Solid-State Circuits*, vol. SC-9, no. 5, pp. 256–268, October, 1974.

[14] P. Chatterjee, W. Hunter, T. Holloway, and Y. Lin, "The Impact of Scaling Laws on the Choice of n-channel or p-channel for MOS VLSI," *IEEE Electron Device Letters*, vol. 1, no. 10, pp. 220–223, October, 1980.

[15] J. Kao, *Subthreshold Leakage Control Techniques for Low Power Digital Circuits*, Ph.D. Thesis, Massachusetts Institute of Technology, May 2001.

[16] V. De and S. Borkar, "Technology and Design Challenges for Low Power and High Performance," in *Proceedings of the International Symposium on Low-Power Electronics and Design*. 1999, pp. 163–168, IEEE/ACM.

[17] S. Thompson, P. Packan, and M. Bohr, "MOS Scaling: Transistor Challenges for the 21st Century," *Intel Technology Journal*, Q3, 1998.

[18] A. Chandrakasan and R. Brodersen, "Minimizing Power Consumption in Digital CMOS Circuits," *in Proceedings of the IEEE*, pp. 498–523, April 1995.

[19] T. Sakurai and A. Newton, "Alpha-Power Law MOSFET Model and its Applications to CMOS Inverter Delay and Other Formulas," *IEEE Journal of Solid-State Circuits*, vol. 25, no. 2, pp. 584–594, 1990.

[20] R. Krishnamurthy and L. Carley, "Exploring the design space of mixed swing quadrail for low-power digital circuits," *IEEE Transactions on VLSI Systems*, vol. 5, no. 4, pp. 388–400, December 1997.

[21] A. Chandrakasan and R. Brodersen, *Low Power Digital CMOS Design*, Kluwer Academics Publications, 1995.

[22] L. Wei, K. Roy, and V. De, "Low Voltage Low Power CMOS Design Techniques for Deep Submicron ICs," *in Proceedings of the 13th International Conference on VLSI Design*, pp. 24–29, January 2000.

[23] K. Roy, "Leakage Power Reduction in Low-Voltage CMOS Designs," in *Proceedings of the International Conference on Electronics, Circuits and Systems*. 1998, pp. 167–173, IEEE.

[24] R. Gu and M. Elmasry, " Power Dissipation Analysis and Optimization of Deep Submicron CMOS Digital Circuits," *IEEE Journal of Solid-State Circuits*, vol. 31, no. 5, pp. 707–713, May 1996.

[25] M. Johnson, D. Somasekhar, and K. Roy, "Models and Algorithms for Bounds on Leakage in CMOS Circuits," *IEEE Transactions on Computer-Aided Design of Integrated Circuits and Systems*, vol. 18, no. 6, pp. 714–725, 1995.

[26] J. Hatler and F. Najm, "A Gate-Level Leakage Power Reduction Method for Ultra Low-Power CMOS Circuits," in *Proceedings of the IEEE Custom Integrated Circuits Conference*, 1997, pp. 475–478.

[27] Y. Ye, S. Borkar, and V. De, "A New Technique for Standby Leakage Reduction in High-Performance Circuits," in *Proceedings of the 1998 Symposium on VLSI Circuits*, June 1998, pp. 40–41.

[28] Z. Chen, L. Wei, and K. Roy, "Estimation of Standby Leakage Power in CMOS Circuits Considering Accurate Modeling of Transistor Stacks," in *Proceedings of the International Symposium on Low-Power Electronics and Design*, August 1998, pp. 239–244.

[29] M. Johnson, D. Somasekhar, and K. Roy, "Leakage Control with Efficient Use of Transistor Stacks in Single Threshold CMOS," *in Proceedings of the 36th Design Automation Conference*, pp. 442–445, June 1999.

[30] J. Rabaey, *Digital Integrated Circuits*, Prentice Hall, 1996.

[31] T. Kuroda, T. Fujita, S. Mita, T. Nagamatsu, S. Yoshioka, K. Suzuki, F. Sano, M. Norishima, M. Murota, M. Kako, M. Kinugawa, M. Kakumu, and T. Sakurai, "A 0.9V 150MHz 10mW $4mm^2$ 2-D Discrete Cosine Transform Core Processor with Variable Threshold Voltage Scheme," *IEEE Journal of Solid-State Circuits*, vol. 31, no. 11, pp. 1770–1779, 1996.

[32] T. Kuroda, T. Fujita, T. Nagamatu, S. Yoshioka, T. Sei, K. Matsuo, Y. Hamura, T. Mori, M. Murota, M. Kakumu, and T. Sakurai, "A High-Speed Low-Power 0.3μm CMOS Gate Array with Variable Threshold Voltage (VT) Scheme," in *Proceedings of the International Custom Integrated Circuits Conference*, 1996, pp. 53–56.

[33] T. Inukai, T. Hiramoto, and T. Sakurai, "Variable threshold voltage CMOS (VTCMOS) in series connected circuits," in *Proceedings of the International Symposium on Low-Power Electronics and Design*, August 2001, pp. 201–206.

[34] I. Hyunsik, T. Inukai, H. Gomyo, T. Hiramoto, and T. Sakurai, "VTCMOS characteristics and its optimum conditions predicted by a compact analytical model," in *in Proceedings of the International Symposium on Low-Power Electronics and Design*, 2001, pp. 123–128.

[35] T. Kuroda, T. Fujita, S. Mita, T. Mori, K. Matsuo, and M. Kakumu, "Substrate Noise Influence on Circuit Performance in Variable Threshold-Voltage Scheme," in *Proceedings of the International Symposium on Low-Power Electronics and Design*, 1996, pp. 309–312.

[36] T. Andoh, A. Furukawa, and T. Kunio, "Design Methodology for Low Voltage MOSFETs," *in Proceedings of the International Electron Devices Meeting*, pp. 79–82, 1994.

[37] F. Assaderaghi, D. Sinitsky, S. Parke, J. Bokor, P. Ko, and C. Hu, "A Dynamic Threshold Voltage MOSFET (DTMOS) for Ultra-Low Voltage Operation," *in Proceedings of the International Electron Devices Meeting*, pp. 809–812, 1994.

[38] W. Lo, S. Chang, C. Chang, and T. Cao, "Impacts of Gate Structure on Dynamic Threshold SOI nMOSFETs," *IEEE Electron Device Letters*, pp. 497–499, August 2002.

[39] T. McPherson et al., "760MHz G6 S/390 Microprocessor Exploiting Multiple Vt and Copper Interconnects," *International Solid-State Circuits Conference Digest of Technical Papers*, pp. 96–97, 2000.

[40] T. Yamashita et al., "A 450MHz 64b RISC Processor using Multiple Threshold Voltage CMOS," *International Solid-State Circuits Conference Digest of Technical Papers*, pp. 414–415, 2000.

[41] S. Tyagi et al., "A 130nm Generation Logic Technology Featuring 70nm Transistors, Dual Vt Transistors and 6 layers of Cu Interconnects," *in Proceedings of the International Electron Devices Meeting*, pp. 567–570, 2000.

[42] Z. Chen, J. Burr, J. Shott, and J. Plummer, "Optimization of Quarter Micron MOSFETs For Low Voltage / Low Power Applications," *in Proceedings of the International Electron Devices Meeting*, pp. 63–66, 1995.

[43] K. Young et al., "A 0.13μm CMOS Technology with 193nm Lithography and Cu/Low-k for High Performance Applications," *in Proceedings of the International Electron Devices Meeting*, pp. 563–566, 2000.

[44] M. Mehrotra et al., "60nm Gate Length Dual-Vt CMOS for High Performance Applications," *in Symposium on VLSI Technology Digest of Technical Papers*, pp. 124–125, 2002.

[45] L. Su et al., "A High-Performance Sub-0.25um CMOS Technology with Multiple Thresholds and Copper Interconnects," *in Symposium on VLSI Technology Digest of Technical Papers*, pp. 18–19, 1998.

[46] Z. Chen, C. Diaz, J. Plummer, M. Cao, and W. Greene, "0.18μm Dual Vt MOSFET Process and Energy-Delay Measurement," *in Proceedings of the International Electron Devices Meeting*, pp. 851–853, 1996.

[47] S. Huang et al., "High Performance 50nm CMOS Devices for Microprocessor and Embedded Processor Core Applications," *in Proceedings of the International Electron Devices Meeting*, pp. 237–240, 2001.

[48] N. Sirisantana, L. Wei, and K. Roy, "High-Performance Low-Power CMOS Circuits Using Multiple Channel Length and Multiple Oxide Thickness ," *in Proceedings of the International Conference on Computer Design*, pp. 227–232, 2000.

[49] S. Parihar et al., "A High Density 0.10μm CMOS Technology Using Low K Dielectric and Copper Interconnect," *in Proceedings of the International Electron Devices Meeting*, pp. 249–252, 2001.

[50] S. Thompson et al., "An Enhanced 130 nm Generation Logic Technology Featuring 60 nm Transistors Optimized for High Performance and Low Power at 0.7 - 1.4V," *in Proceedings of the International Electron Devices Meeting*, pp. 257–260, 2001.

[51] K. Imai et al., "CMOS Device Optimization for System-on-a-chip Applications," *in Proceedings of the International Electron Devices Meeting*, pp. 455–458, 2000.

[52] M. Chang et al., "A Highly Manufacturable $0.25\mu m$ Multiple-Vt Dual Gate Oxide CMOS Process for Logic/Embedded IC Foundry Technology," *in Symposium on VLSI Technology Digest of Technical Papers*, pp. 150–151, 1998.

[53] Y. Taur, S. Wind, Y. Mii, Y. Lii, D. Moy, K. Jenkins, C. Chen, P. Coane, D. Klaus, J. Bucchignano, M. Rosenfield, M. Thomson, and M. Polcari, "High Performance $0.1\mu m$ CMOS Devices with 1.5V Power Supply," *in Proceedings of the International Electron Devices Meeting*, pp. 127–130, 1993.

[54] S. Mutah, T. Douseki, Y. Matsuya, T. Aoki, S. Shigematsu, and J. Yamada, "1-V Power Supply High-Speed Digital Circuit Technology with Multi-Threshold Voltage CMOS," *IEEE Journal of Solid-State Circuits*, vol. 30, no. 8, pp. 847–853, August 1995.

[55] S. Shigematsu, S. Mutoh, and Y. Matsuya, "Power Management Techniques for 1-V LSIs using Embedded Processors," in *Proceedings of the IEEE Custom Integrated Circuits Conference*, 1996, pp. 111–114.

[56] S. Shigematsu, S. Mutah, Y. Matsuya, Y. Tanabe, and J. Yamada, "A 1-V High-Speed MTCMOS Circuit Scheme for Power-Down Application Circuits," *IEEE Journal of Solid-State Circuits*, vol. 32, no. 6, pp. 861–869, 1997.

[57] M. Stan, "CMOS Circuits with Subvolt Supply Voltages," *IEEE Design & Test of Computers*, vol. 19, no. 2, pp. 34–43, March-April 2002.

[58] M. Stan, "Low-Threshold CMOS Circuits with Low Standby Current," in *Proceedings of the International Symposium on Low-Power Electronics and Design*, 1998, pp. 97–99.

[59] H. Kawaguchi, K. Nose, and T. Sakurai, "A CMOS Scheme for 0.5V Supply Voltage with Pico-Ampere Standby Current," *International Solid-State Circuits Conference Digest of Technical Papers*, pp. 192–193, 1998.

[60] T. Inukai, M. Takamiya, K. Nose, H. Kawaguchi, T. Hiramoto, and T. Sakurai, "Boosted Gate MOS (BGMOS): Device/Circuit Cooperation Scheme to Achieve Leakage-Free Giga-Scale Integration," in *Proceedings of the International Custom Integrated Circuits Conference*, 2000, pp. 409–412.

[61] L. Wei, Z. Chen, K. Roy, M. Johnson, and V. De, "Design and Optimization of Dual-Threshold Circuits for Low-Voltage Low-Power Applications," *IEEE Transactions on VLSI Systems*, vol. 7, no. 1, pp. 16–24, March 1999.

[62] S. Sirichotiyakul, T. Edwards, O. Chanhee, Z. Jingyan, A. Dharchoudhury, R. Panda, and D. Blaauw, "Stand-by Power Minimization through Simultaneous Threshold Voltage Selection and Circuit Sizing," in *Proceedings of the 36th Design Automation Conference*, 1999, pp. 436–441.

[63] D. Takashima, S. Watanabe, H. Nakano, Y. Oowaki, and K. Ohuchi, "Standby/active Mode Logic for Sub-1V Operating ULSI Memory," *IEEE Journal of Solid-State Circuits*, vol. 29, no. 4, pp. 441–447, April 1994.

[64] H. Yoo, "Dual-v_t Self-Timed CMOS Logic for Low Subthreshold Current Multigigabit Synchronous DRAM," *IEEE Transactions on Circuits and Systems-II: Analog and Digital Signal Processing*, vol. 45, no. 9, pp. 1263–1271, 1998.

[65] N. Shibata, H. Morimura, and M. Watanabe, "A 1-V, 10-MHz, 3.5-mW, 1-Mb MTC-MOS SRAM with Charge-Recycling Input/Output Buffers," *IEEE Journal of Solid-State Circuits*, vol. 34, no. 6, pp. 866–877, June 1999.

[66] L. Wei, Z. Chen, and K. Roy, "Mixed-v_{th} (MVT) CMOS Circuit Design Methodology for Low Power Applications," *in Proceedings of the 36th Design Automation Conference*, pp. 430–435, June 1999.

[67] V. Sundararajan and K. Parhi, "Low Power Synthesis of Dual Threshold Voltage CMOS VLSI Circuits," *in Proceedings of the IEEE International Symposium on Low Power Electronics and Design*, pp. 139–144, 1999.

[68] Q. Wang and S. Vrudhula, "Algorithms for Minimizing Standby Power in Deep Submicrometer, Dual-v_t CMOS Circuits," *IEEE Journal of Solid-State Circuits*, vol. 21, no. 3, pp. 306–318, March 2002.

[69] S. Sirichotiyakul, T. Edwards, C. Oh, R. Panda, and D. Blaauw, "Duet: An Accurate Leakage Estimation and Optimization Tool for Dual-Vt Circuits," *IEEE Transactions on VLSI Systems - Special Issue on Low Power Electronics and Design*, vol. 10, no. 2, pp. 79–90, April 2002.

[70] L. Wei, K. Roy, and C. Koh, "Power Minimization by Simultaneous Dual-v_{th} Assignment and Gate-sizing," *in Proceedings of the IEEE Custom Integrated Circuits Conference*, 2000, pp. 413–416.

[71] T. Karnik, Y. Ye, J. Tschanz, L. Wei, S. Burns, V. Govindarajulu, V. De, and S. Borkar, "Total Power Optimization By Simultaneous Dual-Vt Allocation and Device Sizing in High Performance Microprocessors," *in Proceedings of the 39th Design Automation Conference*, 2002, pp. 486–491.

[72] K. Roy, L. Wei, and Z. Chen, "Multiple-VDD Multiple-Vth CMOS (MVCMOS) for Low Power Applications," *in Proceedings of the IEEE International on Circuits and Systems*, pp. 366–370, 1999.

[73] M. Khellah and M. Elmasry, "Power Minimization of High-Performance Submicron CMOS Circuits Using a Dual-Vdd/Dual-Vth (DVDV) Approach," *in Proceedings of the IEEE International Symposium on Low Power Electronics and Design*, pp. 106–108, 1999.

Further Reading

The $(V_{dd} - V_{th})$ Design Space

■ D. Liu and C. Svensson, "Trading Speed for Low Power by Choice of Supply and Threshold Voltages," *IEEE Journal of Solid-State Circuits*, vol.28, no.1, pp. 10-17, January 1993.

■ Z. Chen, J. Burr, J. Schott, and J. Plummer, "Optimization of Quarter Micron MOSFETs for Low Voltage/Low Power Applications," *in Proc. IEEE International Electron Devices Meeting*, pp. 63-66, December 1995.

■ Z. Chen, C. Diaz, J. Plummer, M.Cao, and W. Greene, "0.18μm Dual Vt MOSFET Process and Energy-Delay Measurement," *in Proc. IEEE International Electron Devices Meeting*, pp. 851-854, December 1995.

■ H. Oyamatsu, M. Kinugawa, and M. Kakumu, "Design Methodology of Deep Submicron CMOS Devices for 1V Operation," *IEICE Trans. Electron.*, vol. E79-C, no. 12, pp. 1720-1725, December 1996.

■ Q. Wang and S. Vrudhula, "An Investigation Of Power Delay Trade-offs On Powerpc," *in Proc. ACM/IEEE Design Automation Conference*, pp. 425-428, June 1997.

■ D. Franks, P. Solomon, S. Reynolds, and J. Shin, "Supply and Threshold Voltage Optimization for Low Power Design," *in Proc. IEEE International Symposium on Low Power Electronics and Design*, pp. 317-322, August 1997.

■ R. Gonzalez, B. Gordon, and M. Horowitz, "Supply and Threshold Voltage Scaling for Low Power CMOS," *IEEE Journal of Solid-State Circuits*, vol.32, no.8, pp. 1210-1216, August 1997.

■ S. Thompson, I. Young, J. Gearson, and M. Bohr, "Dual Threshold Voltages and Substrate Bias: Keys to High Peformance, Low Power, 0.1μm Logic Designs," *in Symposium on VLSI Technology Digest of Technical Papers*, pp. 69-70, June 1997.

■ K. Chen and C. Hu, "Performance and V_{dd} Scaling in Deep Submicrometer CMOS," *IEEE Journal of Solid-State Circuits*, vol.33, no.10, pp. 1586-1589, October 1998.

■ Q. Wang and S. Vrudhula, "An Investigation of Power Delay Trade-offs for Dual V_t CMOS Circuits," *in Proc. IEEE International Conference on Computer Design*, pp. 556-562, October 1999.

- K. Nose and T. Sakurai, "Optimization of V_{DD} and V_{TH} for Low-Power and High-Speed Applications," *in Proc. Asia and South Pacific - Design Automation Conference*, pp. 469-474, January 2000.

- K. Bernstein, M. Bhushan, and N. Rohrer, "On the Selection of the Optimal Threshold Voltages for Deep Submicron CMOS Technologies," *IBM Microelectronics Journal*, pp. 29-31, First Quarter, 2001.

- A. Wang, A. Chandrakasan, and S.Kosonocky, "Optimal Supply and Threshold Voltage Scaling for Subthreshold CMOS Circuits," *in Proc. of IEEE Computer Society Annual Symposium on VLSI*, pp. 5-9, April 2002.

- M. Vujkovic and C. Sechen, "Optimal Power-Delay Curve Generation for Standard Cell ICs," *in Proc. IEEE International Conference on Computer-Aided Design*, pp. 387-394, November 2002.

Leakage Current Characterization, Modeling, and Estimation

- P. Antognetti, D. Caviglia, and E. Profumo, "CAD Model for Threshold and Subthreshold Conduction in MOSFET's," *IEEE Journal of Solid-State Circuits*, vol. SC-17, no.3, pp. 454-458, June 1982.

- R. Gu and M. Elmasry, "Power Dissipation Analysis and Optimization of Deep Submicron CMOS Digital Circuits," *IEEE Journal of Solid-State Circuits*, vol.31, no.5, pp. 707-713, May 1996.

- S. Thompson, I. Young, J. Gearson, and M. Bohr, "Dual Threshold Voltages and Substrate Bias: Keys to High Peformance, Low Power, $0.1\mu m$ Logic Designs," *in Symposium on VLSI Technology Digest of Technical Papers*, pp. 69-70, June 1997.

- Z. Chen, L. Wei, M. Johnson, and K. Roy, "Estimation of Standby Leakage Power in CMOS Circuits Considering Accurate Modeling of Transistor Stacks," *in Proc. IEEE International Symposium on Low Power Electronics and Design*, pp. 239-244, August 1998.

- S. Bobba and I. Hajj, "Maximum Leakage Power Estimation for CMOS Circuits," *in Proc. IEEE Workshop on Low-Power Design*, pp. 116-124, March 1999.

- M. Johnson, D. Somasekhar, and K. Roy, "Models and Algorithms for Bounds on Leakage in CMOS Circuits," *IEEE Transactions on Computer-Aided Design of Integrated Circuits and Systems*, vol.18, no.6, pp. 714-725, June 1999.

- A. Keshavarzi, K. Roy, and C. Hawkins, "Intrinsic Leakage in Deep Submicron CMOS ICs – Measurement-Based Test Solutions," *IEEE Transactions on VLSI Systems*, vol.8, no.6, pp. 717-723, December 2000.

- D. Sylvester and H.Kaul, "Future Performance Challenges in Nanometer Design," *in Proc. ACM/IEEE Design Automation Conference*, pp. 3-8, June 2001.

- D. Sylvester and H.Kaul, "Power-Driven Challenges in Nanometer Design," *IEEE Design & Test of Computers*, vol.18, no.6, pp. 12-21, November-December 2001.

- Z. Chen, L. Wei, A. Keshavarzi, and K. Roy, "I_{DDQ} Testing for Deep-Submicron ICs: Challenges and Solutions," *IEEE Design & Test of Computers*, vol.19, no.2, pp. 24-33, March-April 2002.

■ Y. Lin, C. Wu, C. Chang, R. Yang, W. Chen, J.Liaw, and C. Diaz, "Leakage Scaling in Deep Submicron CMOS for SoC," *IEEE Transactions on Electron Devices*, vol.49, no.6, June 2002.

■ A. Keshavarzi, J. Tschanz, S. Narendra, V. De, K. Roy, C. Hawkins, W. Daasch, and M. Sachdev, "Leakage and Process Variation Effects in Current Testing on Future CMOS Circuits," *IEEE Design & Test of Computers*, vol.19, no.5, pp. 36-43, September-October 2002.

■ L. Cao, "Circuit Power Estimation using Pattern Recognition Techniques," *in Proc. IEEE/ACM International Conference on Computer-Aided Design*, pp. 412-417, November 2002.

■ R. Kumar and C. Ravikumar, "Leakage Power Estimation for Deep Submicron Circuits in an ASIC Design Environment," *in Proc. IEEE International Conference on VLSI Design*, pp. 45-50, January 2002.

■ G. Yeap, "Leakage Current in Low Standby Power and High Performance Devices: Trends and Challenges," *in Proc. International Symposium on Physical Design*, pp. 22-27, April 2002.

■ A. Ferre and J. Figueras, "Leakage Power Bounds in CMOS Digital Technologies," *IEEE Transactions on Computer-Aided Design of Integrated Circuits and Systems*, vol.21, no.6, pp. 731-738, June 2002.

■ W. Liao, J. Basile, and L. He, "Leakage Power Modeling and Reduction with Data Retention," *in Proc. IEEE/ACM International Conference on Computer-Aided Design*, pp. 714-719, November 2002.

■ K. Roy, S. Mukhopadhyay, and H. Mahmoodi-Meimand, "Leakage Current Mechanisms and Leakage Reduction Techniques in Deep-Submicrometer CMOS Circuits," *Proceedings of the IEEE*, vol.91, no.2, pp. 305-327, February 2003.

■ H. Su, F. Liu, A. Devgan, E. Acar, and S. Nassif, "Full Chip Leakage Estimation Considering Power Supply and Temperature," *in Proc. IEEE International Symposium on Low Power Electronics and Design*, August 2003. (To appear)

■ R. Rao, A. Srivastava, D. Sylvester, and D. Blaauw, "Statistical Estimation of Leakage Current Considering Inter- and Intra-Die Process Variation," *in Proc. IEEE International Symposium on Low Power Electronics and Design*, August 2003. (To appear)

■ E. Acar, A. Devgan, R. Rao, F. Liu, and S. Nassif, "Leakage and Leakage Sensitivity Computation for Combinational Circuits," *in Proc. IEEE International Symposium on Low Power Electronics and Design*, August 2003. (To appear)

■ R. Rao, J. Burns, and R. Brown, "Efficient Techniques for Gate Leakage Estimation," *in Proc. IEEE International Symposium on Low Power Electronics and Design*, August 2003. (To appear)

■ S. Mukhopadhyay and K. Roy, "Modeling and Estimation of Total Leakage Current in Nano-scaled CMOS Devices Considering the Effect of Parameter Variation," *in Proc. IEEE International Symposium on Low Power Electronics and Design*, August 2003. (To appear)

Body Biasing to Minimize Leakage

- T. Kuroda, T. Fujita, T. Nagamatu, S. Yoshioka, T. Sei, K. Matsuo, Y. Hamura, T. Mori, M. Murota, M. Kakumu, and T. Sakurai, "A High-Speed Low-Power 0.3μm CMOS Gate Array with Variable Threshold Voltage (VT) Scheme," *in Proc. IEEE Custom Integrated Circuits Conference*, pp. 53-56, May 1996.

- K. Suzuki, S. Mita, T. Fujita, F. Yamane, F. Sano, A. Chiba, Y. Watanabe, K. Matsuda, T. Maeda, and T. Kuroda, "A 300MIPS/W RISC Core Processor with Variable Supply-Voltage Scheme in Variable Threshold-Voltage CMOS," *in Proc. IEEE Custom Integrated Circuits Conference*, pp. 587-590, May 1997.

- Y. Oowaki, M. Noguchi, S.Takagi, D. Takashima, M. Ono, Y. Matsunaga, K. Sunouchi, H. Kawaguchiya, S. Matsuda, M. Kamoshida, T. Fuse, S. Watanabe, A. Toriumi, S. Manabe, and A. Hojo, "A Sub-0.1μm Circuit Design with Substrate-over-Biasing," *in IEEE International Solid-State Circuits Conference Digest of Technical Papers*, pp. 88-89, February 1998.

- A. Keshavarzi, S. Narendra, S. Borkar, C. Hawkins, K. Roy, and V. De, "Technology Scaling Behavior of Optimum Reverse Body Bias for Standby Leakage Power Reduction in CMOS IC's" *in Proc. IEEE International Symposium on Low Power Design and Electronics*, pp. 252-254, August 1999.

- T. Hiramoto and M. Takamiya, "Low Power and Low Voltage MOSFETs with Variable Threshold Voltage Controlled by Back-Bias," *IEICE Trans. Electron.*, vol. E83-C, no.2, pp. 161-169, February 2000.

- T. Kuroda, T. Fujita, F. Hatori, and T. Sakurai, "Variable Threshold-Voltage CMOS Technology," *IEICE Trans. Electron.*, vol. E83-C, no. 11, pp. 1705-1715, November 2000.

- K. Nose, M. Hirabayashi, H. Kawaguchi, S. Lee, and T. Sakurai, "V_{TH}-hopping Scheme for 82% Power Saving in Low-voltage Processors," *in Proc. IEEE Custom Integarted Circuits Conference*, pp. 93-96, May 2001.

- S. Huang, C. Wann, Y. Huang, C. Lin, T. Schafbauer, S. Cheng, Y. Cheng, D. Vietzke, M. Eller, C. Lin, Q. Ye, N. Rovedo, S. Biesemans, P. Nguyen, R. Dennard, and B. Chen, "Scalability and Biasing Strategy for CMOS with Active Well Bias," *in Symposium on VLSI Technology Digest of Technical Papers*, pp. 107-108, June 2001.

- S. Kosonocky, M. Immediato, P.Cottrell, T. Hook, R. Mann, and J. Brown, "Enhanced Multi-Threshold (MTCMOS) Circuits Using Variable Bias," *in Proc. IEEE International Symposium on Low Power Electronics and Design*, pp. 165-169, August 2001.

- A. Keshavarzi, S. Ma, S. Narendra, B. Bloechel, K. Mistry, T. Ghani, S. Borkar, and V. De, "Effectiveness of Reverse Body Bias for Leakage Control in Scaled Dual Vt CMOS ICs," *in Proc. IEEE International Symposium on Low Power Electronics and Design*, pp. 207-212, August 2001.

- C. Kim and K. Roy, "Dynamic V_{TH} Scaling Scheme for Active Leakage Power Reduction," *in Proc. Design, Automation and Test in Europe Conference and Exhibition*, pp. 163-167, March 2002.

- M. Miyazaki, J. Kao, and A. Chandrakasan, "A 175mV Multiply-Accumulate Unit using an Adaptive Supply Voltage and Body Bias (ASB) Architecture," *IEEE Journal of Solid-State Circuits*, vol. 37, no.11, pp. 1545-1554, November 2002.

- A. Keshavarzi, S. Narendra, B. Bloechel, S. Borkar, and V. De, "Forward Body Bias for Microprocessors in 130nm Technology Generation and Beyond," *in Symposium on VLSI Circuits Digest of Technical Papers*, pp. 312-315, June 2002.

- S. Martin, K. Flautner, T. Mudge, and D. Blaauw, "Combined Dynamic Voltage Scaling and Adaptive Body Biasing for Low Power Microprocessors under Dynamic Workloads," *in Proc. IEEE/ACM International Conference on Computer Aided Design*, pp. 721-725, November 2002.

- J. Tschanz, J. Kao, S. Narendra, R. Nair, D. Antoniadis, A. Chandrakasan, and V. De, "Adaptive Body Bias for Reducing Impacts of Die-to-Die and Within-Die Parameter Variations on Microprocessor Frequency and Leakage," *IEEE Journal of Solid-State Circuits*, vol.37, no.11, pp. 1396-1402, November 2002.

- T. Sakurai, "Minimizing Power across Multiple Technology and Design Levels," *in Proc. IEEE/ACM International Conference on Computer Aided Design*, pp. 24-27, November 2002.

- J. Tschanz, S. Narendra, Y. Ye, B. Bloechel, S. Borkar, and V. De, "Dynamic Sleep Transistor and Body Bias for Active Leakage Power Control of Microprocessors," *in International Solid-State Circuits Conference Digest of Technical Papers*, pp. 102-103, February 2003.

- C. Neau and K. Roy, "Optimal Body Bias Selection for Leakage Improvement and Process Compensation Over Different Technology Generations," *in Proc. IEEE International Symposium on Low Power Electronics and Design*, August 2003. (To appear)

Scaling of Leakage Current

- S. Asai and Y. Wada, "Technology Challenges for Integration Near and Below $0.1\mu m$," *Proceedings of the IEEE*, vol.85, no.4, pp. 505-520, April 1997.

- S. Thompson, I. Young, J. Gearson, and M. Bohr, "Dual Threshold Voltages and Substrate Bias: Keys to High Peformance, Low Power, $0.1\mu m$ Logic Designs," *in Symposium on VLSI Technology Digest of Technical Papers*, pp. 69-70, June 1997.

- A. Dancy and A. Chandrakasan, "Techniques for Aggressive Supply Voltage Scaling and Efficient Regulation," *in Proc. IEEE Custom Integrated Circuits Conference*, pp. 579-586, May 1997.

- M. Bohr and Y. El-Mansy, "Technology for Advanced High-Performnace Microprocessors," *IEEE Transactions on Electron Devices*, vol.45, no.3, pp. 620-625, March 1998.

- H. Wong, D. Franks, P. Solomon, C.Wann, and J. Welser, "Nanoscale CMOS," *Proceedings of the IEEE*, vol.87, v.4, pp. 537-570, April 1999.

- Y. Taur, "The Incredible Shrinking Transistor," *IEEE Spectrum*, vol.36, no.7, pp. 25-29, July 1999.

- A. Keshavarzi, S. Narendra, S. Borkar, C. Hawkins, K. Roy, and V. De, "Technology Scaling Behavior of Optimum Reverse Body Bias for Standby Leakage Power Reduction in CMOS IC's" *in Proc. IEEE International Symposium on Low Power Design and Electronics*, pp. 252-254, August 1999.

- T. Sakurai, "Design Challenges for $0.1\mu m$ and Beyond," *in Proc. Asia and South Pacific Design Automation Conference*, pp. 553-558, January 2000.

- S. Huang, C. Wann, Y. Huang, C. Lin, T. Schafbauer, S. Cheng, Y. Cheng, D. Vietzke, M. Eller, C. Lin, Q. Ye, N. Rovedo, S. Biesemans, P. Nguyen, R. Dennard, and B. Chen, "Scalability and Biasing Strategy for CMOS with Active Well Bias," *in Symposium on VLSI Technology Digest of Technical Papers*, pp. 107-108, June 2001.

- S. Narendra, S. Borkar, V. De, D. Antoniadis, and A. Chandrakasan, "Scaling of Stack Effect and its Application for Leakage Reduction," *in Proc. IEEE International Symposium on Low Power Electronics and Design*, pp. 195-200, August 2001.

- A. Keshavarzi, S. Ma, S. Narendra, B. Bloechel, K. Mistry, T. Ghani, S. Borkar, and V. De, "Effectiveness of Reverse Body Bias for Leakage Control in Scaled Dual Vt CMOS ICs," *in Proc. IEEE International Symposium on Low Power Electronics and Design*, pp. 207-212, August 2001.

- D. Sylvester and H.Kaul, "Power-Driven Challenges in Nanometer Design," *IEEE Design & Test of Computers*, vol.18, no.6, pp. 12-21, November-December 2001.

- Y. Lin, C. Wu, C. Chang, R. Yang, W. Chen, J.Liaw, and C. Diaz, "Leakage Scaling in Deep Submicron CMOS for SoC," *IEEE Transactions on Electron Devices*, vol.49, no.6, June 2002.

- A. Keshavarzi, J. Tschanz, S. Narendra, V. De, K. Roy, C. Hawkins, W. Daasch, and M. Sachdev, "Leakage and Process Variation Effects in Current Testing on Future CMOS Circuits," *IEEE Design & Test of Computers*, vol.19, no.5, pp. 36-43, September-October 2002.

- G. Yeap, "Leakage Current in Low Standby Power and High Performance Devices: Trends and Challenges," *in Proc. International Symposium on Physical Design*, pp. 22-27, April 2002.

- G. Sery, S. Borkar, and V. De, "Life Is CMOS: Why Chase the Life After?," *in Proc. ACM/IEEE Design Automation Conference*, pp. 78-83, June 2002.

- C. Neau and K. Roy, "Optimal Body Bias Selection for Leakage Improvement and Process Compensation Over Different Technology Generations," *in Proc. IEEE International Symposium on Low Power Electronics and Design*, August 2003 (To appear).

- B. Chatterjee, S. Hsu, R. Krishnamurthy, and S. Borkar, "Effectiveness and Scaling Trends of Leakage Control Techniques for Sub-130nm CMOS Technologies," *in Proc. IEEE International Symposium on Low Power Electronics and Design*, August 2003 (To appear).

Gate Leakage

- K. Brown, L. Wang, X. Tang, and J. Meindel, "A Circuit-Level Perspective of the Optimum Gate Oxide Thickness," *IEEE Transactions on Electron Devices*, vol.48, no.8, pp. 1800-1810, August 2001.

- F. Hamzaoglu and M. Stan, "Circuit-Level Techniques to Control Gate Leakage for sub-100nm CMOS," *in Proc. IEEE International Symposium on Low Power Electronics and Design*, pp. 60-63, August 2002.

- D. Lee, W. Kwong, D. Blaauw, and D. Sylvester, "Analysis and Minimization Techniques for Total Leakage Considering Gate Oxide Leakage," *in Proc. ACM/IEEE Design Automation Conference*, pp. 175-180, June 2003.

- A. Agarwal and K. Roy, "A Noise Tolerant Cache Design to Reduce Gate and Subthreshold Leakage in the Nanometer Regime," *in Proc. IEEE International Symposium on Low Power Electronics and Design*, August 2003 (To appear).
- R. Rao, J. Burns, and R. Brown, "Efficient Techniques for Gate Leakage Estimation," *in Proc. IEEE International Symposium on Low Power Electronics and Design*, August 2003 (To appear).
- R. Rao, J. Burns, and R. Brown, "Circuit Techniques for Gate and Subthreshold Leakage Minimization in Future CMOS Technologies," *in Proc. European Solid-State Circuits Conference*, September 2003 (To appear).
- M. Drazdziulis and P. Larsson-Edefors, "A Gate Leakage Reduction Strategy for Future CMOS Circuits," *in Proc. European Solid-State Circuits Conference*, September 2003 (To appear).

Subthreshold Leakage Reduction in Digital Systems

- W. Lee et al., "A 1-V Programmable DSP for Wireless Communications," *IEEE Journal of Solid-State Circuits*, v32, no.11, pp. 1766-1776, November 1997.
- C. Akrout, "A 480-MHz RISC Microprocessor in a 0.12-μm L_{eff} CMOS Technology with Copper Interconnects," *IEEE Journal of Solid-State Circuits*, vol.33, no.11, pp. 1609-1616, November 1998.
- T. McPherson et al., "760MHz G6/390 Microprocessor Exploiting Multiple Vt and Copper Interconnects," *in IEEE International Solid-State Circuits Conference Digest of Technical Papers*, pp. 96-97, February 2000.
- T. Yamashita at al., "A 450MHz 64b RISC Processor using Multiple Threshold Voltage CMOS," *in IEEE International Solid-State Circuits Conference Digest of Technical Papers*, pp. 414-415, February 2000.
- D. Duarte, Y. Tsai, N. Vijaykrishnan, and M. Irwin, "Evaluating Run-Time Techniques for Leakage Power Reduction," *in Proc. International Conference on VLSI Design*, pp. 31-38, January 2002.
- J. Tschanz, Y. Ye, L. Wei, V. Govindarajulu, N. Borkar, S. Burns, S. Burns, T. Karnik, S. Borkar, and V. De, "Design Optimization of a High Performance Microprocessor Using Combinations of Dual-V_T Allocation and Transistor Sizing," *in Symposium on VLSI Circuits Digest of Technical Papers*, pp. 218-219, June 2002.
- L. Clark, S. Demmons, N. Deutscher, and F. Ricci, "Standby Power Management for a 0.18μm Microprocessor," *in Proc. IEEE International Symposium on Low Power Electronics and Design*, pp. 7-12, August 2002.
- S. Vangal et al., "5-GHz 32-bit Integer Execution Core in 130-nm Dual-V_T CMOS," *IEEE Journal of Solid-State Circuits*, vol.37, no.11, pp. 1421-1432, November 2002.
- R. Rao, J. Burns, and R. Brown, "Circuit Techniques for Gate and Subthreshold Leakage Minimization in Future CMOS Technologies," *in Proc. European Solid-State Circuits Conference*, September 2003 (To appear).
- B. Calhoun and A. Chandrakasan, "Standby Voltage Scaling for Reduced Power," *in Proc. IEEE Custom Integrated Circuits Conference*, September 2003 (To appear).

Subthreshold Leakage Reduction in Memory

- M. Powell, S. Yang, B. Falsafi, K. Roy, and T. Vijaykumar, "Reducing Leakage in a High-Performance Deep-Submicron Instruction Cache," *IEEE Transactions on VLSI Systems*, vol.9, no.1, pp. 77-89, February 2001.

- H. Hanson, M. Hrishikesh, V. Agarwal, S. Keckler, and D. Burger, "Static Energy Reduction Techniques for Microprocessor Caches," *in Proc. International Conference on Computer Design*, pp. 276-283, September 2001.

- K. Flautner, N. Kim, S. Martin, D. Blaauw, and T. Mudge, "Drowsy Caches: Simple Techniques for Reducing Leakage Power," *in Proc. International Symposium on Computer Architecture*, pp. 148-157, May 2002.

- A. Agrawal, H. Li, and K. Roy, "DRG-Cache: A Data Retention Gated-Ground Cache for Low Power," *in Proc. ACM/IEEE Design Automation Conference*, pp. 473-478, June 2002.

- N. Azizi, A. Moshovos, and F. Najm, "Low-Leakage Asymmetric-Cell SRAM," *in Proc. IEEE International Symposium on Low Power Electronics and Design*, pp. 48-51, August 2002.

- A. Agarwal and K. Roy, "A Noise Tolerant Cache Design to Reduce Gate and Subthreshold Leakage in the Nanometer Regime," *in Proc. IEEE International Symposium on Low Power Electronics and Design*, August 2003 (To appear).

Chapter 3

EMBEDDED MTCMOS COMBINATIONAL CIRCUITS

3.1. Introduction

Several computer-aided methodologies have been proposed in the literature to optimally design MTCMOS circuits. Some of those methodologies employed high-V_{th} switch sleep transistors, whereas others have involved embedded high-V_{th} transistors or gates. This chapter presents CAD methods to optimally design MTCMOS circuits with embedded high-V_{th} transistors or gates.

3.2. Basic Concept

The basic concept of embedded multi-threshold voltages circuits is to use both high-V_{th} cells and low-V_{th} cells flexibly in a logic block.

The concept relies on the observation that a circuit's overall performance is often limited by a few critical paths. The transistors and gates along these critical paths are set to a low-V_{th}, whereas their transistor sizes are fixed. By assigning a few transistors on the critical paths of the circuit to a low-V_{th}, the overall circuit performance can be improved significantly, while the leakage current is kept within bounds. An example of the path distribution of a synthesized circuit is illustrated in Figure 3.1; the circuit's performance can be increased by 19% by speeding up only 15% of the total paths in the circuit. This approach was used in the PowerPC 750 [1] [2].

To further clarify this concept, Figure 3.2 shows the flip-flop (FF)-to-FF path delay distribution of a logic block. Generally, these distributions are used in the logic design process to evaluate the performance of a logic circuit. The operating frequency of the logic block is limited by the maximum path delay value. In other words, when the distributions are moved to the left by chang-

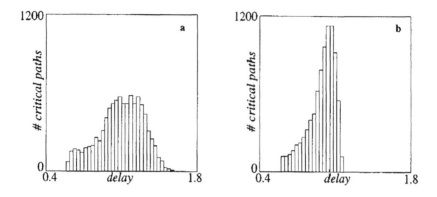

Figure 3.1.   Path delay distribution of a circuit before and after size optimization [3]

ing the circuit configuration to reduce the delay, the operating frequency is increased. For example, when the maximum path delay of all-high-V_{th} cells is 1.0 (a. u.) and the ratio of the average delay of the low-V_{th} cells to that of the high-V_{th} cells is 0.8:1.0, the objective is to reduce the maximum path delay to 0.8 by combining the low-V_{th} cells.

Figure 3.2 demonstrates that using all-low-V_{th} cells reduces the maximum path delay to 0.8. Since the operating frequency is determined by the maximum (the worst) delay, the distribution moves too far to the left. A gradated modulation scheme, presented in [4] and briefly mentioned in Section 3.4, gradually changes the ratio of the low-V_{th} cells according to the path delay. To achieve the objective delay of 0.8, it is necessary to change all the cells included in the maximum delay path to low-V_{th} cells. However, if all the cells in a path with a 0.9 delay are changed to low-V_{th} ones, the path delay becomes 0.72. So, to achieve just a 0.8 delay, only half of the cells in such paths need to be low-V_{th} cells. Gradated modulation attempts to move the delay of all the violated paths to just 0.8, so that the number of low-V_{th} cells is minimized. This scheme produces a steep distribution such as the delta function in Figure 3.2. Note that the maximum path delay after gradated modulation is exactly the same as that of all-low-V_{th} cells.

In conclusion, the objective is to utilize the multi-threshold technology to transform the bell-shaped path delay distribution into a steep distribution that is similar to the delta function. This guarantees that performance is maintained, while the leakage power is reduced by employing high-V_{th} gates.

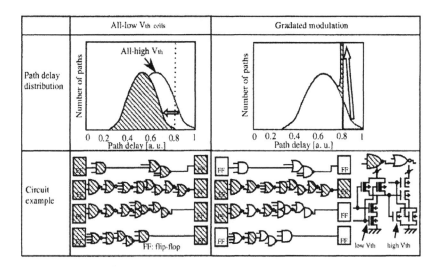

Figure 3.2.   The Concept of gradated modulation [4]

To devise design methodologies that transform the delay path distributions into a delta function, and to consequently minimize leakage power, two issues must be considered carefully: (1) accurate circuit delay models must be generated through static timing analysis, and (2) the leakage current must be accurately modeled.

(1) Modeling Gate Delays

Sizing or changing the threshold voltage of a gate (G) affects not only its delay and power, but also the delay and power of the neighboring gates.

For example, sizing a gate G changes the load capacitance of its fanin gates GI and the input transition time of its fanout gates GO (Figure 3.3. Due to the change of the load capacitance of GI, the output transition time of GI is different, which in turn, changes the delay and power of all the fanout gates of GI; GIO. If the threshold voltage of gate G is also changed, the delay and power of GO will change. To summarize, sizing or changing the threshold voltage of G influences the delay and power of G, GI, GO, and GIO which are the neighbors of gate G.

In general, the propagation delay through node x in gate G, denoted as $t_p(x)$, defines how quickly the output responds to a change in the input. The

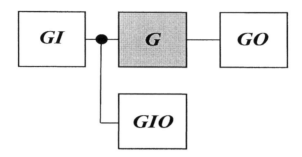

Figure 3.3. Gate G and neighboring gates

propagation delay of a path K_j is the sum of the propagation delays $t_p(i)$ of each node i along this path. It can be expressed as $t_p(K_j) = \sum t_p(i)$.

Elmore Delay Model

In order to model the gate delay, the delay of the transistors is modeled [6]. An n-input NAND gate (see Figure 3.4) is first examined. Each transistor has an equivalent resistance R_j, and each internal node in the n-input NAND gate has a capacitance of C_j ($j=1...n$). The equivalent RC network of the PullDown Network (PDN) is shown in Figure 3.5. The worst case occurs when all C_js are discharged simultaneously.

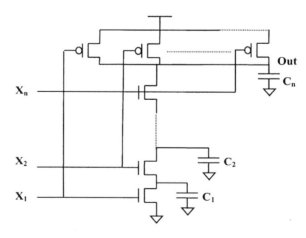

Figure 3.4. n-input NAND gate

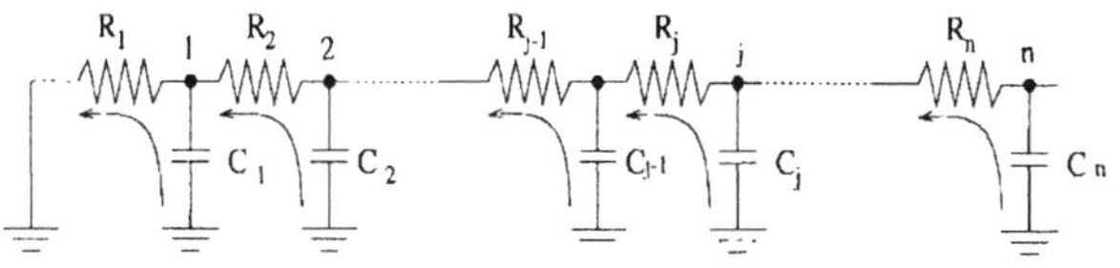

Figure 3.5. Equivalent PDN of n-input NAND gate

Based on the Elmore delay model [5], the worst case high-to-low propagation delay t_{PHL} of the PDN is given by

$$t_{PHL} = 0.69 \sum_{j=1}^{n} \left(C_j \sum_{k=1}^{j} R_k \right) \tag{3.1}$$

The capacitance of each internal node j in the n-input NAND gate is given as

$$C_j = 2C_{dN} \tag{3.2}$$

where C_{dN} is the diffusion capacitance of an NMOS device. The capacitance of the gate at the output is given by

$$C_n = F_O(C_{gP} + C_{gN}) + F_I C_{dP} + C_{dN} + F_O C_{int} \tag{3.3}$$

where C_{dP} is the diffusion capacitance of a PMOS device, C_{gP} and C_{gN} are the gate capacitance of the PMOS and NMOS devices, respectively, C_{int} represents the interconnect capacitance per fanout, F_O is the fanout of the gate, and F_I is the number of fanins. If it is assumed that the NMOS devices have similar *on* resistances, the high-to-low propagation delay can be approximated as

$$t_{PHL} = 0.69[R_N C_{dN} F_I(F_I - 1) + F_I R_N C_n] \tag{3.4}$$

For the PMOS pullup network, the worst case occurs when only one PMOS device is *on*. The low-to-high propagation delay can be expressed as

$$t_{PLH} = 0.69 R_P C_n \tag{3.5}$$

Finally, the worst case propagation of the CMOS NAND gate is written as

$$t_p = (t_{PHL} + t_{PLH})/2 \tag{3.6}$$

Following a similar procedure, the worst case propagation delay of the other gates can be computed. By using the expression derived for the delay at node x, three timing parameters are defined for a gate G located in a circuit: (1) the arrival time, (2) the departure time or the required time as it is times referred to, and (3) the slack.

(1) **The arrival time** $T_a(x)$ is the propagation delay of each fanin path of node x. Among all the fan-in paths, there exists a path(s) which has a maximum propagation delay time $T_{max}(x)$ where

$$T_{max}(x) = \max_{i \in all\, fanins} \{T_a(x)[i]\} \tag{3.7}$$

(2) **The departure time** $T_l(x)$ of node x is defined as the latest time the signal has to arrive at the output of gate G,

$$T_l(x) = T_{max}(x) + t_p(x) \tag{3.8}$$

The path which determines the maximum speed of the circuit is called the critical path. There may be more than one critical path in a single circuit.

(3) **The slack time** of the node $T_\delta(x)$ represents the amount by which a gate can be slowed down without affecting the circuit performance. For the nodes in the critical path, the slack is zero. For a primary output (PO),

$$T_\delta(PO) = T_{critical} - T_l(f_{anin}(PO)) \tag{3.9}$$

For any other node x, $T_\delta(x)$ can be expressed as

$$T_\delta(x) = \min_{\forall y = fanout(x)} \{(T_\delta(y) + T_{max}(y) - T_l(x))\} \tag{3.10}$$

where f_{anin} and $f_{anout}(x)$ are the fanin and fanout nodes of node x, respectively. Equation (3.10) ensures that the propagation delay of the path(s) through x is never greater than the critical delay. By definition, for each node in a single low-threshold circuit, its slack (T_δ) is greater or equal to zero. Increasing the threshold voltage of a node can result in a higher propagation delay and departure time of that node. Therefore, the slack time will increase. Whether a node should be assigned to a high V_{th} depends on whether its slack time is still positive, when the node's threshold voltage is changed from low to high.

If the slack time is still positive, then the node can be assigned a high V_{th}. On the other hand, the threshold voltage of the nodes located on the critical path is not changed, because its slack is zero. The set of gates that has the minimum slack value constitutes the critical path of the circuit.

(2) Leakage Current Estimation

A critical issue in leakage current optimization is obtaining an accurate and meaningful metric for the leakage current of a circuit that can be efficiently calculated and used in an optimization engine.

It has been illustrated in Chapter 2 that leakage power is composed of several mechanisms. The subthreshold leakage current is the component with the highest value. The junction leakage (reverse currents in the source/drain junction diodes with the bulk) is two to three orders smaller than I_{sub} and is ignored. Similarly, the reverse junction current between the well and the bulk is ignored, because the reverse junction current is significantly smaller and is usually not a target for optimization at the circuit level. Therefore, the subthreshold current is the focus in leakage current modeling [1].

From the BSIM2 MOS transistor model, the subthreshold leakage current is modelled as

$$I_{sub} = A exp(\frac{q}{nkT}(V_g - V_s - V_{th_0} - \gamma V_s + \eta V_{ds}))[1 - exp(\frac{-qV_{ds}}{kT})] \quad (3.11)$$

where

$$A = \mu_0 C_{ox} \frac{W_{eff}}{L_{eff}} (\frac{kT}{q})^2 e^{1.8}$$

C_{ox} is the gate oxide capacitance per unit area, μ_0 is the zero-bias mobility, n is the subthreshold swing coefficient of the transistor, V_{th_0} is the zero-bias threshold voltage, γ is the linearized body effect coefficient, and η is the DIBL coefficient.

In a CMOS logic gate, the leakage current depends strongly on the relative position of the *on* and *off* devices in the transistor network of the gate, and the logic state of the gate inputs (state probabilities of the gate).

Position of Devices

If the transistors are connected in a parallel configuration (e.g., NMOS devices in a NOR gate) and are both turned *off*, then V_{ds} and V_s are similar for each transistor. The leakage current of each transistor can be calculated separately and then summed up. This, however, is not true for transistors connected in series (e.g., NMOS devices in a NAND gate). When the series transistors are off, the quiescent subthreshold current through each transistor must be equal. By equating the leakage current of the upper device to that of the lower transistor, V_{ds_2} is obtained ($V_{s_1} \ll V_{dd}$) as follows:

$$V_{ds_2} = \frac{nkT}{q(1 + 2\eta + \gamma)} ln(\frac{A_1}{A_2} e^{q\eta V_{dd}/nkT} + 1) \qquad (3.12)$$

The voltage of the ith transistor in terms of the $(i\text{-}1)$th transistor is given as

$$V_{ds_i} = \frac{nkT}{q(1+\gamma)} ln(1 + \frac{A_{i-1}}{A_i}(1 - e^{(-q/kT)V_{ds_{i-1}}}) \qquad (3.13)$$

Equation (3.13) can be used iteratively to find V_{ds_i} for each transistor. Using Equation (3.11), the quiescent leakage (I_{ds_q}) for any transistor in the stack is calculated .

In a large circuit, the leakage current is computed by the following steps:

- Transistors that are on are considered to be short-circuit.

- The leakage current of a transistor in parallel with an on transistor is ignored.

- For the resultant network, V_{ds} is estimated for the remaining transistors using Equations (3.12) and (3.13).

- The magnitude of the leakage current is then computed.

Finally, the total leakage power $P_{leakage}$ can be determined as the sum of $V_{ds_q} \times I_{ds_q}$ over all the transistors such that

$$P_{leakage} = \sum_i I_{ds_{q_i}} V_{ds_{q_i}} \qquad (3.14)$$

However, this stack-based model ignores the voltage drops across the on transistors in the stack such as in [6] [7] and [8]. These procedures can result in significant errors as revealed in [3]. Therefore, a method that uses a nonlinear simulation with accurate leakage models, similar to those in [9] [10], must be used.

Unfortunately, the calculation of the leakage current is complicated by the highly nonlinear behavior of the drain current of a device with respect to source/drain voltages. Several simple models for subthreshold operations which are efficient for circuit simulation have been proposed in [9] and [11]. Nevertheless, an accurate SPICE simulation by using these nonlinear models is still very expensive, and is not feasible for the repeated evaluations of large circuits in an optimization framework. Furthermore, leakage current of a circuit is highly dependent on the state of the circuit. Therefore the state probabilities of a circuit must be considered to save CPU time, which will be examined next.

State Probabilities

Figure 3.6 reflects the leakage current for all states of a three-input NAND gate. For this gate, the highest leakage current is 99 times greater than the lowest one, clearly indicating a strong dependence of the leakage on the circuit state. When the leakage current of the circuit as a whole is considered, the correlation between the states of the gates must also be considered.

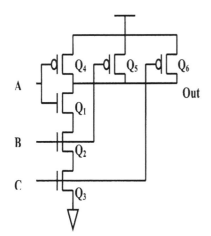

State (ABC)	Leakage Current (nA)	Leaking Transistors
000	0.095	Q_1, Q_2, Q_3
001	0.195	Q_1, Q_2
010	0.195	Q_1, Q_3
011	1.874	Q_1
100	0.184	Q_2, Q_3
101	1.220	Q_2
110	1.140	Q_3
111	9.410	Q_4, Q_5, Q_6

Figure 3.6. Leakage current in a 3-input NAND gate [3]

It is necessary to simulate a substantial portion of the gates' states to obtain an accurate average leakage of each gate. Since the number of gates in a circuit is usually very large, it would require an extremely large number of random global circuit vectors to sufficiently cover the states of the individual DCCs, and for their average leakage to converge. The complexity of the average leakage estimation is reduced through a sequence of steps, which are presented in a top-down manner as follows:

- A probabilistic approach eliminates the need to do simulations over all 2^n input combinations (where n is the number of circuit inputs).

- A small subset of all the possible states is evaluated for leakage. This approach is based on the notion of dominant-leakage states which will be explained next.

Calculation of State Probabilities

The methods presented in [6] and [12] use statistical simulations to measure the average leakage of an entire circuit. These statistical methods are based on Monte Carlo experiments, where, in each iteration, a circuit state which is randomly chosen is applied, and the leakage of the entire circuit is computed. The iteration is conducted until the average leakage of the circuit, computed over all the applied vectors converges. However, a leakage-optimization tool relies on an accurate estimation of the leakage current rather than estimates for the total circuit leakage.

For this reason, the probabilistic approach is more effective than statistical simulations for optimization purposes. The probabilistic approach previously described accounts for the leakage of every state, provided such leakage is significant (i.e., dominant). Although fewer states are explored than by statistical simulations, the confidence in the accuracy of the estimated average leakage is high.

Dominant Leakage States

In experiments using SPICE simulations of several gates in all the possible states, the leakage of a gate is significantly less in some states than in others. A state with more than one transistor off in a path from V_{dd} to $ground$ (a $V_{dd}\text{-}ground$ path) is far less leaky than a state with only one transistor off in any $V_{dd}\text{-}ground$ path. The latter state is called a dominant leakage states; a set of dominant leakage states is usually very small compared with a set of all the possible states. The key idea is to ignore the leakage of insignificant (non-dominant) states in the average leakage calculation without the loss of accuracy. For example, SPICE simulations of a three-input NAND gate in Figure 3.6 show that the exact average leakage is 1.78925 nA, if equal probabilities are assumed for all the states [3]. The set of dominant leakage states for this gate is $D = \{011, 101, 110, 111\}$. If only these four states are considered, the average leakage is 1.7055 nA, only 4.68% less than the exact average. Note that such accuracy is obtained by simulating only four out of the eight possible states. This trade-off becomes even more attractive for circuits with a large number of inputs.

The following section discusses the formulization of the Power Minimization Problem.

3.3. The Power Minimization Problem

In general, the power minimization problem in any of the proposed algorithms can be stated as follows:

Given: (1) a random logic network of N static CMOS gates, (2) the critical delay of the circuit that is less than or equal to the maximum specified delay T_{max} as usually dictated by the clock frequency, (3) a device technology, and (4) activity profiles at each input node

Determine: (1) the supply voltage V_{dd}, (2) the threshold voltage of each MOSFET V_{th}, and (3) the channel width of each MOSFET W.

Such that: (1) the static leakage and dynamic power are minimized, and (2) the area is within bounds. The leakage power to be minimized in this power minimization problem is the subthreshold leakage power.

Therefore, the power minimization problem can be cast as a constrained optimization problem. Its solution solves the following question: **What are the sizes and threshold voltages for the transistors, as well as the supply voltage, that achieve the best performance/power trade-off for the circuit without exceeding a given total area?**

This is the general formulation of the power minimization problem, but there are some special cases for the problem:

- The supply voltage V_{dd} takes a single value.
- The threshold voltage of a device can take one of three possibilities:

  (1) A **continuous** domain parameter, for example, architectures that employ the variable threshold voltage approach (VTCMOS) [13] by utilizing a triple well structure and biasing the substrate of each device.

  (2) Multiple **discrete** values.

  (3) The actual high-V_{th} and low-V_{th} values of a device are fixed.

  Condition (3) is the most popular, and it simplifies the threshold voltage selection problem to one with a discrete domain with only two choices: a high-V_{th} or a low-V_{th}.

- In general, each transistor can be individually assigned a V_{th} value (high or low). However, V_{th} can be assigned in a stack-based or gate-based optimization by selecting the V_{th} of a predefined group of transistors, where a group consists of transistors in the same stack or in the same gate. Stack-based or gate-based optimizations are preferable, if there is a manufacturing process limitation on assigning different values to closely-spaced transistors. This occurs when the transistors are in a stack, and their channels are too close to each other, making it difficult to achieve distinct channel doping. Therefore, it is hard to obtain different thresholds for the transistors in the stack. Gate-based optimization is also more suitable for a standard cell design methodology.

- The area can be excluded as a design constraint in the optimization problem. This reduces the problem from a 3-D problem, where speed, power and area are optimized, to a 2-D problem, where only speed and power are the parameters that must be optimized.

- Some techniques define power in the optimization problem as the summation of the dynamic and leakage power, whereas others are only concerned with minimizing the leakage power.

The Formulation

Let P be the total number of paths in the circuit, and let the jth path be numbered K_{j_1}, K_{j_2},..., $K_{j_{n_j}}$, where n_j is the number of gates in the jth path. If the static and dynamic power dissipation of the ith gate are denoted as P_{static} and $P_{dynamic}$, respectively, and its delay as t_{d_i}, the nonlinear optimization problem can be stated as

$$Minimize : \sum_{i=1}^{N} (P_{static_i} + P_{dynamic_i}) \qquad (3.15)$$

subject to

$$\sum_{i=K_{j_1}}^{K_{j_{n_j}}} t_{d_i} \leq T_{max}, j = 1, ..., P \qquad (3.16)$$

The dynamic power in a CMOS circuit is due to the charging and discharging of the load capacitances, as well as the internal-node capacitances. This can be evaluated as follows:

$$P_{dynamic} = P_{dynamic_o} + P_{dynamic_j} \qquad (3.17)$$

and

$$P_{dynamic} = \frac{1}{2}f(V_{dd}{}^2 \sum_i (\alpha_i C_{L_i}) + V_{dd} \sum_i \sum_j (\alpha_{ij} C_{ij} V_{ij})) \qquad (3.18)$$

and

$$P_{static} = \sum_{i=1}^{N} I_{static_i} V_{dd} \qquad (3.19)$$

$P_{dynamic_o}$ and $P_{dynamic_j}$ are the dynamic power dissipations due to load capacitances and internal-node capacitances, respectively, f is the clock frequency, i represents gate i, j denotes the jth internal node in a gate, V_{ij} is the voltage swing of the jth internal node of gate i, α_i and α_{ij} are the switching activities at gate i and at the jth internal node of gate i, respectively, and C_{L_i} and C_{ij} are the load capacitance and the jth internal-node capacitance of gate i, respectively. The switching activity can be determined by a Monte-Carlo-based statistical method [14].

In general, in order to obtain a better trade-off between the performance and leakage of a design, the assignment of low V_{th} and high V_{th} transistors must be performed at the same time as the transistor sizes are being adjusted. If, in a well-balanced circuit, the V_{th} of a transistor on the critical path is lowered, while the transistor size is fixed, the path will become unnecessarily fast, thereby making the sizes suboptimal. Also, a lower V_{th} causes the formation of the channel to occur earlier during the input transition due to the earlier onset of the strong inversion with a reduced V_{th}. This results in an increase of approximately 8 to 10% of the average gate capacitance of the transistor gate as seen by the input driver [3]. As a result, when a transistor is assigned a low V_{th}, the increase of its gate capacitance can adversely affect the performance of the other paths loaded by this transistor's gate node. Therefore, setting a transistor to a low-V_{th}, without subsequently adjusting the transistor sizes in the circuit, can actually degrade the performance of the circuit and increase leakage. Therefore, the transistor sizes must be adjusted simultaneously with the assignment of values to obtain an optimal solution.

Several algorithms to minimize the leakage power will now be examined.

3.4.   Algorithms

In the literature, the proposed algorithms to minimize leakage power while achieving the desired speed can be divided into three categories according to how the threshold voltages are assigned, (1) a transistor-level assignment, (2) a gate-level assignment, or (3) a mix between gate and transistor-level assignment.

Simultaneous V_{dd}, V_{th}, and W Optimization at the Transistor-Level [15] [16]

Delays are first assigned to all the gates of a circuit without violating the cycle time constraint. Then, the optimal supply voltage, threshold voltage, and transistor width of each gate are determined so as to minimize power consumption. The power to be minimized is comprised of dynamic and leakage power. It should be pointed out that the estimation of the standby power in [15] [16] does not take into account the signal probabilities.

Design and Optimization of Dual-Threshold Circuits at the Gate-Level [14] [17]

The first step in this algorithm is to initialize the circuit with a single low-V_{th}. The next step is to assign a high-V_{th} to some transistors on noncritical paths under performance constraints. This is performed by back-tracing the slack of each node level by level. The pseudo-code of the algorithm is shown in Figure 3.7

```
High-Vth Assignment {
    For each node x {
    Calculate tp(x), Tl(x), and Tδ(x) for high-Vth
    if Tδ(x) ≥ 0
        Assign high-Vth to x
        Assign tp(x), Tl(x), and Tδ(x) for high-Vth to x
    else
        Keep tp(x), Tl(x), and Tδ(x) for initial low-Vth for x
    }
```

Figure 3.7. Pseudo-code of High-V_{th} Assignment Algorithm

The method described in [14] [17] consists of assigning high threshold voltages to transistors in noncritical paths, and low threshold voltages to transistors in critical paths. The threshold voltage assignment is done at the logic gate level; that is, all the transistors within a gate are either at a low-V_{th} or at a high-V_{th}. If an initial configuration in which all the gates are at a low-V_{th} is assumed, each gate is examined in turn and replaced by a high-V_{th} gate, if it does not decrease the minimum slack over all the nodes. Two drawbacks associated with [14] [17] are: (1) the greedy nature of the heuristic precludes many possible configurations in which sets of transistors on and off critical paths can be

assigned different threshold voltages to reduce the standby current without degrading the performance, and (2) the possibility of different transistors within
a logic gate having different threshold voltages is also not considered.

Mixed-V_{th} (MVT) CMOS Circuit Design [17]

For mixed-V_{th} CMOS circuits, the transistors within a gate have different
threshold voltages with certain process constraints. There are two types of
mixed-V_{th} CMOS circuit schemes. For the Type 1 scheme (MVT1), there is
no mixed V_{th} in the PMOS pullup or NMOS pulldown trees. Figure 3.8 shows
an example circuit in MVT1. In the type 2 scheme (MVT2), a mixed V_{th} is allowed anywhere except in the series connected transistors. Figure 3.8 provides
an example of a circuit in MVT2.

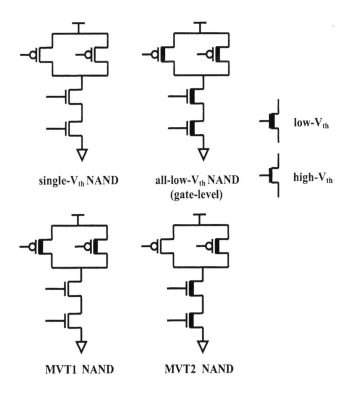

Figure 3.8. Mixed-V_{th} schemes [17]

The transistors in a stack are assigned the same threshold voltage because
of the process considerations that are explained in Section 3.3. The algorithm

in [17], which is used for assigning high and low V_{th}, is an extension of the gate-level algorithm in [14]. Dual threshold voltages are assigned to the transistors under performance constraints. All the transistors in the circuit are assumed to initially have a low-V_{th}. The circuit is forward-traced level by level from the primary inputs to calculate the departure time of each gate. Next, the circuit is back-traced level by level from the primary outputs to explore every gate G.

The pullup tree slack (T_δ^p), pulldown tree slack (T_δ^n), and the slack of each transistor within G is calculated. For the MVT1 scheme, if all the transistors in the pullup (pulldown) tree of gate G satisfy the requirements that Δt_d are not larger than their slack values, the pullup (pulldown) tree can be assigned a high-V_{th}. Δt_d is the difference between the high-V_{th} delay and the low-V_{th} delay.

In the MVT2 scheme, for each transistor of gate G, if it is not a series connected transistor and its Δt_d is not larger than its slack value, this transistor can be assigned a high-V_{th}. For the series-connected transistors, if the Δt_d of all the transistors in the series are not larger than their slack values, a high-V_{th} is assigned to all the transistors in the series. Otherwise, a low-V_{th} is maintained. After the threshold voltage is assigned to each transistor within gate G, the propagation delay of each transistor within G is updated. Then, the departure time of gate G is recalculated.

Low-Power Synthesis of Dual-Threshold Voltage CMOS VLSI Circuits at the Gate-level [18]

In [14] [17], a simple backward breadth first search heuristics to identify the subset of gates that can be switched to a high-V_{th}. In [18], a near optimal strategy for the solution of the problem presented in [14] [17] is devised. Based on delay balancing by buffer insertions, the approach is a tool that captures all the slack in the circuit. The delay buffers are fictitious entities whose sole purpose is to model the slack present in the circuit, and will therefore not contribute to the circuit's power dissipation. The fictitious buffers are referred to as SDF (Specific Delay Fictitious) buffers. The objective is then to employ these SDF-displacements to identify that particular delay balanced configuration which identifies a new threshold voltage for all the logic gates, while providing a maximum reduction in leakage power in the standby mode and maintaining the critical path in the active mode. Similar to the constraints of the techniques previously described, the constraints governing the minimization of the leakage power ensure that after the SDF-displacement, each wire in the circuit has a non negative delay value for the SDF buffers on that wire.

Power Minimization by Simultaneous Dual-V_{th} Assignment and Gate-Sizing at the Gate-Level [19]

The technique presented in [19] is based on the TILOS (TImed LOgic Synthesis) [20] algorithm which was originally used to minimize the area subject to delay constraints. The TILOS algorithm can be described as follows:

- All the transistors in the circuit are minimum sized.

- A static timing analysis is used to find the critical delay path to determine the maximum speed of the circuit. In order to reduce the critical delay with a minimal increase in the area, TILOS finds the sensitivity of the delay with respect to the area for all the transistors in the critical path, and up-sizes the transistor with the largest sensitivity.

- The delay of each transistor is then updated by static timing analysis.

- The process is iterated until the delay constraint is satisfied.

The TILOS algorithm is modified to minimize the total power and critical delay by simultaneous gate-sizing and dual-V_{th} assignment. The proposed technique is a gate-level V_{th} assignment, indicating that every gate can have only a single V_{th} value.

The technique starts from the minimal size and single high-V_{th} circuit, which corresponds to the minimal dynamic and leakage power. To optimize the circuit, the critical path is identified and a gate is set that can reduce the critical delay the best, but with a minimal increase in the total power. That gate is modified by either increasing its size or reducing the threshold voltage. The circuit is then updated using static timing analysis. The process is iterated until changing the gate size and threshold voltage cannot further improve the critical delay.

Gradated Modulation: Multi-Threshold-Voltage Design at the Gate-Level [4]

The pseudo-code of the gradated modulation procedure for combining the two kinds of V_{th} cells is shown in Figure 3.9. The key point of this procedure is determining which cell should be assigned the low-V_{th} in order to minimize the total number of low-V_{th} cells as mentioned earlier in Section 3.2. The cell is selected according to an evaluation function $EF(cell)$ as follows:

```
Procedure Gradated Modulation {
    read net-list;
    change all cells to high-Vth;
    calculate delay and slack;
    while (violated path exists) {
        select the minimum slack path;
        for each cell in the selected path {
            calculate evaluation value of each cell;
        }
        while (slack of the pathis negative) {
            select the cell that has the highest evaluated value;
            change the selected cell to low-Vth;
            re-calculate delay and slack;
        }
    }
}
```

Figure 3.9.   Pseudo-code of Gradated Modulation procedure

$$EF(cell) = \frac{t_{pd}(cell)}{I_{sub}(cell)} \times N_{vp}(cell) \tag{3.20}$$

where t_{pd} and I_{sub} are the delay and the leakage current of the cell, and N_{vp} is the number of violated paths that include the cell as illustrated in Figure 3.2. This evaluation function is constructed according to the following two viewpoints:

- to select such cells that the reduction of the delay is large and the increase of leakage current is small.

- to select such cells that decrease the delay of several violated paths at once.

Figure 3.10 exemplifies a simple circuit example to explain the second viewpoint. In Figure 3.10, there are two violated paths. To solve the delay-violation, three low-V_{th} cells are required for each path. In Figure 3.10(a), only three cells are changed to low-V_{th} ones, because the selected cells are included in both violated paths. However, in Figure 3.10(b), six cells are changed to low-V_{th} ones. The leakage current of the circuit in Figure 3.10(b) is about twice that of (a). Owing to the evaluation factor N_{vp} in Equation (3.20), the cells are selected according to circuit (a).

Simultaneous Sizing and V_{th} Assignment at the Transistor-Level [3] [21]

In this technique, both the selection and transistor size of the circuit are considered simultaneously. Two techniques are used in this approach:

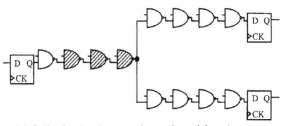

(a) Cell selection base on the evaluated function

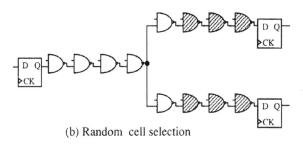

(b) Random cell selection

Figure 3.10. Modulated circuit example [4]

(1) the selection uses leakage and performance sensitivities to accurately determine the impact of changing the transistor's V_{th}, and (2) the V_{th} selection process incrementally adjusts the transistor sizes of the circuit after each change in V_{th}.

An iterative approach is proposed that uses a merit function to evaluate the increase in the total leakage with respect to the performance gain of the whole circuit. In each iteration, the merit function is calculated for all the transistors in the circuit, and the transistor with the best merit is selected and is assigned a low-V_{th}. The circuit sizes are then re-balanced which will be explained next, and the circuit timing and transistor merit is incrementally recalculated; the procedure is then repeated. The merit function is written as

$$Merit(T) = \frac{\Delta I_{sub}(T)}{\Delta D(T)} \tag{3.21}$$

The factor $\Delta I_{sub}(T)$ is the change in leakage of the circuit due to the change in the V_{th} of transistor T. The factor $\Delta D(T)$ captures the impact of lowering the V_{th} for transistor T on all the affected paths in the circuit, which are weighted according to their criticality.

The advantage of this approach is that in each iteration, the transistor which increases the circuit performance the most, relative to its increase in leakage is selected, while taking into account the effect of both the increased drive strength and the increased capacitance on the performance of the circuit as a whole.

Re-Balancing

As explained earlier, a circuit's device size is no longer optimal once the V_{th} of one or more transistors is lowered. Both the speed and gate capacitance of such transistors are changed, affecting all the incident timing paths and increasing the load on nearby transistors. The reduction in delay results in excess area, whereas the increase in capacitance results in undersized devices. By shifting the excess area to undersized regions, performance can be improved without an area penalty. The process of adjusting the device widths, while maintaining the total circuit area, is called *rebalancing*; it is accomplished in two steps: (1) reducing the selected transistor widths and (2) resizing the entire circuit back to its original area.

(1) The first step of rebalancing is *reduction*; the removal of excess area from the devices which are faster than necessary. Lowering the V_{th} of a transistor can easily change its speed by as much as 50% [3]. Such a localized speed increase introduces an imbalance among the paths. Removing the area from the selected devices is the first step toward correcting the imbalance. The set of reduction candidates includes any device sharing a timing path with a V_{th}-changed device, and the V_{th}-changed device itself.

(2) The second step of re-balancing is *resizing*, or optimally distributing excess area in order to reduce the worst circuit delay. A delay/area sensitivity-based size optimization tool for the resizing step [22] balances the delay of all the timing paths, thus minimizing the total circuit area for a given performance. Although the resizing phase initially focuses on only the obviously undersized devices affected during the reduction step, all the devices in the circuit are candidates for resizing, and the excess area is distributed across all the critical timing paths [3].

3.5. Choosing the High-V_{th} Value

The problem between trading off speed for lower leakage power during the standby mode can be resolved by using dual-threshold voltages. A low V_{th} is assigned to the transistors located in the critical path(s) in order to achieve high performance, while high V_{th} is assigned to the transistors in the non-critical paths to reduce the leakage power. However, not all the transistors in the non-critical paths can be assigned a high V_{th}. Otherwise, some non-critical paths may become critical. Whether a transistor can be assigned a high V_{th} or not depends on the value of the high threshold. If a high-V_{th} takes a small value, there will be little difference in the propagation delay between the low-V_{th} and high-V_{th} transistors. As a result, more transistors can be assigned a high-V_{th} without impacting the critical delay. However, the improvement in leakage current savings for each high-V_{th} transistor will be small. Yet, if the high-V_{th} is too large, the leakage current will be greatly reduced. However, fewer transistors can be assigned a high-V_{th}.Consequently, there is an optimal high-V_{th} value that achieves the maximum leakage power reduction without degrading the circuit's performance.

Figure 3.11 reflects the high threshold voltage versus usage ratio of the cell and leakage current of an $0.2\mu m$ 450MHz 64-bit RISC processor [4]. A low-V_{th} of 240mV is chosen to meet the target frequency, and the high-threshold voltage varies from 240mV to 740mV. The 240mV high-V_{th} represents a single-low-V_{th} design.

The usage ratio of high-V_{th} cells decreases, and that of the low-V_{th} cells increases when the high threshold voltage is increased. This occurs because the delay of the paths consisting of the high-V_{th} cells increases, and the high-V_{th} cells can only be used for a limited number of paths. As a result, the leakage current of a high-V_{th} cell decreases, and that of a low-V_{th} cell increases; thus, the total leakage current has a minimum point at a high-V_{th} of 340mV so that the difference is 100mV.

Therefore, a rule of thumb for optimized threshold voltages can be written as [23]

$$(high - V_{th}) - (low - V_{th}) = 0.1V \qquad (3.22)$$

Impact on Well-Optimized Designs

The discussed algorithms in Section 3.4 provide good leakage power savings for many of the circuits. However, the algorithms have difficulty optimizing circuits that are carefully balanced by using post-synthesis optimization

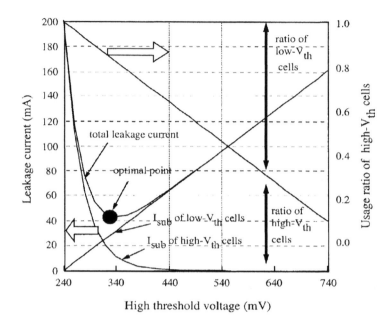

Figure 3.11.   High threshold voltage versus leakage current and usage ratio of cell [4]

techniques such as *transistor sizing*. Figure 3.1 portrayed the path distribution for the same circuit after transistor sizing has been applied. To further
increase the performance of this balanced circuit, requires that transistors on a
large portion of all the paths be assigned a low-V_{th} resulting in a far less favor-
able trade-off between performance and leakage current.

To investigate the impact of well-optimized designs, the gradated modulation scheme to the whole of a large-scale microprocessor, the following is
deduced: Gradated modulation reduces the leakage current by 75% to 90%
compared to using all-low-V_{th} cells. These reductions are achieved without
degrading performance. The usage ratios of the low-V_{th} cells of each block are
also shown in the table in Figure 3.12.

By using the gradated modulation scheme, the ratio of low-V_{th} cells can be
restrained from 1.6% to 14.7%. The differences in the leakage current reduction occur because of the differences in the delay status of these blocks. The
Cache-control unit (CCU), central processor unit (CPU) and floating point execution unit (FPU) are heavily and carefully delay-optimized in a logic-design,
because they are the timing-critical blocks of the microprocessor. However,

Block Name	(a) All-low-V$_{th}$ design			(b) All-high-V$_{th}$ design			(c) Random modulation			
	Delay (ns)	Low-Vth cell #	Leakage power (mW)	Delay (ns)	Low-Vth cell #	Leakage power (mW)	Delay (ns)	Low-Vth cell #	Leakage reduction power (mW)	(%)
CCU	4.1	16315 (100%)	243.7	4.9	0 (0%)	18.9	4.1	1412 (8.6%)	38.3	84.3
CPU	4.5	15247 (100%)	210.7	5.8	0 (0%)	16.3	4.5	2224 (14.6%)	47.4	77.5
FPU	3.6	10210 (100%)	138.4	4.5	0 (0%)	10.7	3.6	1499 (14.7%)	31.6	77.2
BCU	3.8	12151 (100%)	148.8	4.6	0 (0%)	11.5	3.8	194 (1.6%)	13.9	90.2

Figure 3.12. Usage ratios of low-V_{th} cells [4]

the bus control unit (BCU) is not a timing-critical block of the microprocessor. The delay status of each block is visualized by the proposed cell-delay distribution shown in Figure 3.13.

The cell-delay distribution of the BCU is wide compared with those of the other blocks. The number of cells in the delay-violation range is relatively small. Consequently, it can be seen that the delay status of the BCU is not as severe. The distributions are changed after gradated modulation as expected, and all have a steep peak at the target delay. In the low delay region, the distribution after gradated modulation is very similar to that of the all-high-V_{th} cells. In Figure 3.13, the hatched areas represent the low-V_{th} cell population in the gradated modulation. The low-V_{th} cells are used only in the paths with the target delay.

3.6.   Chapter Summary

This chapter presented the most popular techniques to design MTCMOS circuits with embedded dual-V_{th}. It has been shown that the methodology to optimally design such MTCMOS circuits must involve: (1) accurate modeling of the gate delays in the design, and (2) efficient estimation of the leakage current in every gate. The *Power Minimization Problem* is then defined, taking all possible design critria into account. Finally, the kind of algorithm employed to achieve an optimal solution for the *Power Minimization Problem*. In this chapter, several algorithms at the gate- and transistor-level are illustrated, outlining their main features and drawbacks.

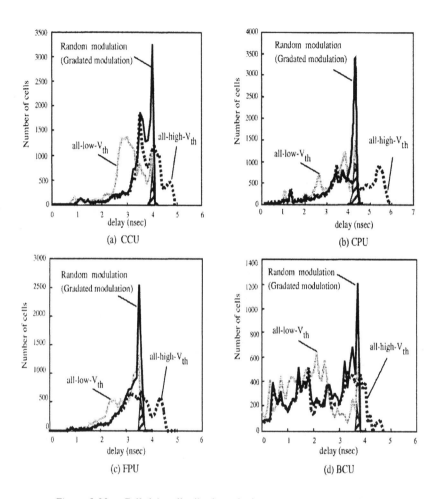

Figure 3.13. Cell-delay distribution of microprocessor components [4]

Notes

1 Modeling of gate leakage has gained large interest in the last couple of
   years. Gate leakage is emerging as an alarming source of power dissipation in nanometer designs.

References

[1] N. Rohrer, C. Akrout, M. Canada, D. Cawthron, B. Davari, R. Floyd, S. Geissler, R. Gold-blatt, R. Houle, P. Kartschoke, D. Kramer, P. McCormick, G. Salem, R. Schulz, L. Su, and L. Whitney, "A 480MHz RISC Microprocessor in a $0.12\mu m$ Leff CMOS Technology with Copper Interconnects," *in International Solid-State Circuits Conference Digest of Technical Papers*, pp. 240–241, 1998.

[2] C. Akrout, M. Canada, D. Cawthron, J. Corr, B. Davari, R. Floyd, S. Geissler, R. Gold-blatt, R. Houle, P. Kartschoke, D. Kramer, P. McCormick, N. Rohrer, G. Salem, R. Schulz, L. Su, and L. Whitney, "A 480MHz RISC Microprocessor in a $0.12\mu m$ Leff CMOS Technology with Copper Interconnects," *IEEE Journal of Solid-State Circuits*, vol. 33, no. 11, pp. 1609–1616, November 1998.

[3] S. Sirichotiyakul, T. Edwards, C. Oh, R. Panda, and D. Blaauw, "Duet: An Accurate Leakage Estimation and Optimization Tool for Dual-Vt Circuits," *IEEE Transactions on VLSI Systems - Special Issue on Low Power Electronics and Design*, vol. 10, no. 2, pp. 79–90, April 2002.

[4] N. Kato, Y. Akita, M. Hiraki, T. Yamashita, T. Shimizu, F. Maki, and K. Yano, "Random Modulation: Multi-Threshold-Voltage Design Methodology in Sub-2-V Power Supply CMOS," *IEICE Transactions Electron.*, vol. E83-C, no. 11, pp. 1747–1754, November 2000.

[5] J. Rabaey, *Digital Integrated Circuits*, Prentice Hall, 1996.

[6] L. Wei, Z. Chen, M. Johnson, K. Roy, and V. De, "Design and Optimization of Low Voltage High Performance Dual Threshold CMOS Circuits," *Proc. of 35th Design Automation Conference*, pp. 489–494, June 1998.

[7] Q. Wang and S. Vrudhula, "Static Power Optimization of Deep Submicron CMOS Circuits for Dual Vt Technology," *in Proceedings of the International Conference on Computer Aided Design (ICCAD)*, pp. 490–496, November 1998.

[8] Z. Chen, L. Wei, and K. Roy, "Estimation of Standby Leakage Power in CMOS Circuits Considering Accurate Modeling of Transistor Stacks," in *Proceedings of the International Symposium on Low-Power Electronics and Design*, August 1998, pp. 239–244.

[9] P. Antognetti, "CAD Model for Threshold and Subthreshold Conduction in MOSFETs," *IEEE Journal of Solid-State Circuits*, vol. SC-17, pp. 454–458, June 1982.

[10] M. Johnson, D. Somasekhar, and K. Roy, "Models and Algorithms for Bounds on Leakage in CMOS Circuits," *IEEE Transactions on Computer-Aided Design of Integrated Circuits and Systems*, vol. 18, no. 6, pp. 714–725, 1995.

[11] M. Chen and J. Ho, "A Three-Parameter-Only MOSFET Subthreshold Current CAD Model Considering Back-gate Bias and Process Variation," *IEEE Transactions on Computer-Aided Design of Integrated Circuits and Systems*, vol. 16, no. 4, pp. 343–352, April 1997.

[12] J. Hatler and F. Najm, "A Gate-Level Leakage Power Reduction Method for Ultra Low-Power CMOS Circuits," in *Proceedings of the IEEE Custom Integrated Circuits Conference*, 1997, pp. 475–478.

[13] T. Kuroda, T. Fujita, S. Mita, T. Nagamatsu, S. Yoshioka, K. Suzuki, F. Sano, M. Norishima, M. Murota, M. Kako, M. Kinugawa, M. Kakumu, and T. Sakurai, "A 0.9V 150MHz 10mW $4mm^2$ 2-D Discrete Cosine Transform Core Processor with Variable Threshold Voltage Scheme," *IEEE Journal of Solid-State Circuits*, vol. 31, no. 11, pp. 1770–1779, Nov. 1996.

[14] L. Wei, Z. Chen, K. Roy, M. Johnson, and V. De, "Design and Optimization of Dual-Threshold Circuits for Low-Voltage Low-Power Applications," *IEEE Transactions on VLSI Systems*, vol. 7, no. 1, pp. 16–24, March 1999.

[15] P. Pant, V. De, and A. Chatterjee, "Device-Circuit Optimization for Minimal Energy and Power Consumption in CMOS Random Logic Networks," in *Proceedings of the Design Automation Conference*, pp. 403–408, 1997.

[16] P. Pant, V. De, and A. Chatterjee, "Simultaneous Power Supply, Threshold Voltage, and Transistor Size Optimization for Low-Power Operation of CMOS Circuits," *IEEE Transactions on VLSI Systems*, vol. 6, no. 4, pp. 538–545, December 1998.

[17] L. Wei, Z. Chen, and K. Roy, "Mixed-v_{th} (MVT) CMOS Circuit Design Methodology for Low Power Applications," in *Proceedings of the 36th Design Automation Conference*, pp. 430–435, June 1999.

[18] V. Sundararajan and K. Parhi, "Low Power Synthesis of Dual Threshold Voltage CMOS VLSI Circuits," in *Proceedings of the IEEE International Symposium on Low Power Electronics and Design*, pp. 139–144, 1999.

[19] L. Wei, K. Roy, and C. Koh, "Power Minimization by Simultaneous Dual-v_{th} Assignment and Gate-sizing," in *Proceedings of the IEEE Custom Integrated Circuits Conference*, 2000, pp. 413–416.

[20] J. Fishburn and A. Dunlop, "TILOS: A Polynomial Programming Approach to Transistor Sizing," in *Proceedings of the International Conference on Computer Aided Design (ICCAD)*, pp. 326–328, 1985.

[21] S. Sirichotiyakul, T. Edwards, O. Chanhee, Z. Jingyan, A. Dharchoudhury, R. Panda, and D. Blaauw, "Stand-by Power Minimization through Simultaneous Threshold Voltage

Selection and Circuit Sizing," in *Proceedings of the 36th Design Automation Conference*, 1999, pp. 436–441.

[22] A. Dharchoudhury, D. Blaauw, T. Noston, S. Pullela, and T. Dunning, "Transistor-level Sizing and Timing Verification of Domino Circuits in the power PC Microprocessor," *in Proceedings of the International Conference on Computer Design (ICCD)*, pp. 143–148, October 1997.

[23] M. Hirabayashi, K. Nose, and T. Sakurai, "Design Methodology and Optimization Strategy for Dual-Vth Scheme using Commercially Available Tools," *in Proceedings of the International Symposium on Low Power Electronics and Design*, pp. 283–286, August 2001.

Further Reading

- M. Khellah and M. Elmasry, "Power Minimization of High-Performance Submicron CMOS Circuits Using a Dual-V_{dd} Dual-V_{th} (DVDV) Approach," *in Proc. IEEE International Symposium on Low Power Electronics and Design*, pp. 106-108, August 1999.

- K. Khouri and N. Jha, "Leakage Power Analysis and Reduction during Behavioral Synthesis," *in Proc. IEEE International Conference on Computer Design*, pp. 561-564, September 2000.

- P. Pant, R. Roy, and A. Chatterjee, "Dual-Threshold Voltage Assignment with Transistor Sizing for Low Power CMOS Circuits," *IEEE Transactions on VLSI Systems*, vol.9, no.2, pp. 390-394, April 2001.

- T. Ishihara and K. Asada, "A System Level Memory Power Optimization Technique Using Multiple Supply and Threshold Voltages," *in Proc. ACM/IEEE Design Automation Conference*, pp. 456-461, June 2001.

- M. Ketkar and S. Sapatnekar, "Standby Power Optimization via Transistor Sizing and Dual Threshold Voltage Assignment," *in Proc. IEEE International on Computer-Aided Design*, pp. 375-378, November 2002.

- J. Kao, S. Narendra, and A. Chandrakasan, "Subthreshold Leakage Modeling and Reduction Techniques," *in Proc. IEEE International on Computer-Aided Design*, pp. 141-148, November 2002.

- T. Karnik, S. Borkar, and V. De, "Sub-90nm Technologies – Challenges and Opportunities for CAD," *in Proc. IEEE International on Computer-Aided Design*, pp. 203-206, November 2002.

- R. Brodersen, M. Horowitz, D. Markovic, B. Nikolic, and V. Stojanovic, "Methods for True Power Minimization," *in Proc. IEEE International on Computer-Aided Design*, pp. 35-42, November 2002.

- K. Khouri and N. Jha, "Leakage Power Analysis and Reduction During Behavioral Synthesis," *IEEE Transactions on VLSI Systems*, vol.10, no.6, pp. 876-885, December 2002.

- A. Srivastava, "Simultaneous Vt Selection and Assignment for Leakage Optimization," *in Proc. IEEE International Symposium on Low Power Electronics and Design*, August 2003. (To appear)

Chapter 4

MTCMOS COMBINATIONAL CIRCUITS USING SLEEP TRANSISTORS

4.1. Introduction

The Multi-Threshold CMOS (MTCMOS) is a very attractive technique to reduce sub-threshold leakage currents during standby modes by utilizing high-V_{th} power switches (sleep transistors). This technology is straightforward to use, because existing designs can be modified to become MTCMOS blocks by simply adding high-V_{th} power supply switches. In addition, circuits can easily be placed in low leakage states at a fine grain level of control. For this reason, and because the *time to market* is critical, this technique gained the attention of the industry. Unlike the embedded MTCMOS designs discussed in Chapter 3, circuits employing high-V_{th} sleep transistors do not require the re-designing of the original low-V_{th} block.

4.2. MTCMOS Design: Overview

The basic MTCMOS structure is depicted in Figure 4.1, where a low-V_{th} (LVT) logic block is gated with high-V_{th} power switches which are controlled by $SLEEP$ signals. When the high-V_{th} transistors are turned *on* ($SLEEP$=1), the low-V_{th} logic gates are connected to virtual *ground* and power, and the switching is performed. During the active mode, the sleep transistor can be realized as a resistor R as shown in Figure 4.1 [1]. The virtual *ground* incorporates a voltage V_X equal to $I \times R$, where I is the current flowing through the sleep transistor.

The voltage drop across R has two effects. First, it reduces the gate's driving capability from V_{dd} to V_{dd}-V_X, and secondly, it causes the threshold voltage of the LVT pulldown devices to increase due to the body effect [2] [3]. Since,

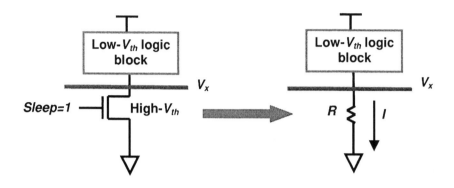

Figure 4.1.   Sleep transistor modeled as resistor

both effects degrade the speed of the circuit, the resistor should be sized small, and consequently, the size of the sleep transistor be made large. However, this comes at the expense of extra area and power. Yet, if the resistor is sized too large (i.e., the sleep transistor is sized small), the circuit speed will be degraded. Therefore, a trade-off exists between achieving sufficient performance and low-power values. This trade-off becomes even more evident in the deep submicron (DSM) regime. In DSM technologies, the supply voltage is scaled down aggressively, causing the resistance of the sleep transistor to increase dramatically, requiring even larger size sleep devices. This will cause leakage and dynamic power to significantly increase in the standby and active modes, respectively. Therefore, an important design criterion is sizing the sleep transistor so that sufficient performance is attained; that is, the current I flowing through the sleep transistor must be sufficient to achieve the required speed.

The worst case scenario takes place if all the gates supported by the sleep transistor are simultaneously switching in time (Figure 4.2). This is the first case, where the sleep transistor exhibits a maximum current such that ($I = I_1 + I_2 + I_3$). In this case, the sleep transistor is sized up to contain the high current. In the second case, the gates are discharging mutually exclusive, and the sleep transistor is sized according to the maximum current of the mutually exclusive discharging gates ($I = \max\{I_1, I_2, I_3\}$). The sleep transistor is much smaller in this case. In the first case, if a current-time graph is constructed of the discharged currents, I_1, I_2 and I_3 will overlap in time, but for the second case, no overlap in time occurs. An intermediate case occurs when the discharged currents *partially* overlap, and the LVT logic blocks have slightly different discharge times.

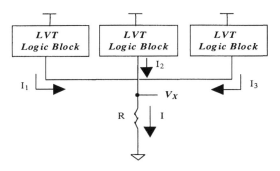

Figure 4.2.   Worst case discharging scenario

Therefore, the proper sizing of the sleep transistor is a key element to efficiently design MTCMOS circuits, and poses as a design challenge in complex MTCMOS circuits, because the critical path input vector is highly dependent on the internal discharge patterns of the core logic. This is attributed to the fact that input vectors that cause more current to flow through the sleep transistor can cause more ground bounce which degardes performance. As a result, it is not sufficient in MTCMOS circuits with sleep transistors to examine only a critical path, as in the pure CMOS case, in order to determine performance. The worst case pattern for a pure CMOS design would, therefore, not translate to the worst case pattern for the MTCMOS implementation of that circuit.

Furthermore, critical paths in MTCMOS circuits not only depend on the input vector, but also on the sleep transistor size. For example, a sleep transistor which is sized large may have a critical path that is long, but there may be few transitions causing minimal virtual ground bounce. However, for that same circuit with a small sleep transistor, large virtual ground bouncing can take place, and a different vector which exercises a shorter path may become critical. Examples that illustrate these points can be found in [4].

Therefore, in MTCMOS circuits with high-V_{th} sleep transistors, all the discharge patterns must be examined in order to accurately size the sleep transistor to determine the worst case MTCMOS input vector for a given sleep transistor size. Due to the exponential explosion in complexity for performing exhaustive simulations, techniques should be devised to achieve a relatively accurate MTCMOS design with a reasonable computation time.

In this chapter, the most popular techniques that attempt to properly size these high-V_{th} sleep transistors are explored; (1) the *Variable Breakpoint* simulator, (2) hierarchical sizing based on mutually exclusive discharge patterns, (3) the average current method, and (4) the distributed sleep transistors technique.

4.3.   Variable Breakpoint Switch Level Simulator [1]

In 1997, Kao et al. developed a *variable breakpoint* simulator to rapidly compute circuit delay as a function of sleep transistor size. This tool identifies potential input vectors that can cause large variations in an MTCMOS circuit, and can also be used to narrow down the vector space to be analyzed with a more detailed simulator such as SPICE.

In order to model the delay of the MTCMOS circuit, the gates are first modeled as inverters. N inverters are assumed to be sharing the same virtual ground rail. If it is assumed that all N gates are simultaneously discharging current (Figure 4.3, the high-to-low propagation delay T_{delay} for a particular gate G ($0<G<N$) can be modeled as

$$T_{delay} \approx \frac{C_L V_{dd}}{2I_j} \tag{4.1}$$

where I_j is the saturation current, and is expressed as

$$I_j = \frac{1}{2}\beta_j(V_{dd} - V_{th} - V_x)^2 \tag{4.2}$$

where V_x is the potential of the virtual ground rail and is equal to

$$V_x = \frac{1}{2}\beta_{total}(V_{dd} - V_{th} - V_x)^2 R \tag{4.3}$$

β_j is the gain factor for MOSFET j and is equal to $\mu_n C_{ox}$(W/L), and $\beta_{total} = \beta_1 + \beta_2 + ... + \beta_n$.

The algorithm behind this simulation tool dynamically adjusts each gate's propagation delay according to the total number of gates switching, since different amounts of current produce different voltage drops across the sleep transistor.

The input and output voltage waveforms for each gate are treated as piecewise linear, and the gates are assumed to begin switching when the input voltage exceeds $V_{dd}/2$. The gates are modeled as time varying (step-wise) current sources discharging their respective load capacitances, resulting in a piecewise

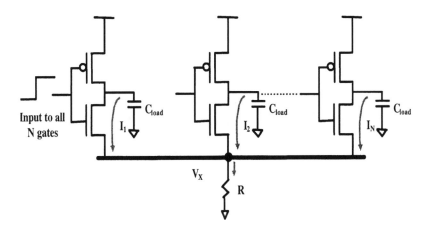

Figure 4.3. N gates sharing the same virtual ground and discharging simultaneously

linear output voltage whose slopes can vary in time. These breakpoints occur whenever a gate in the logic block starts or stops switching, because the delays must be recomputed when the total current flowing through the sleep transistor changes.

To process these breakpoints, the simulator computes a *best guess* for the amount of time required to reach the switching threshold and the amount of time required to finish the switching for each gate. The simulator time steps to the nearest breakpoint, determines if any new elements are switching, and then recomputes the *best guess* for these breakpoints by taking into account the slower or faster gate transitions.

To clarify the idea behind the algorithm for the simulation tool, Figure 4.4 shows the output waveforms as functions of time for three different gates in a MTCMOS circuit. One breakpoint is labelled as t_i , corresponding to the switching threshold of Gate 2, and another is shown as t_{i+1} , corresponding to the time that Gate 1 completes discharging.

Before time t_i, Gate 1 is discharging at a constant slope, and Gate 2 is transitioning from low to high. At the breakpoint t_i , Gate 2 passes the threshold voltage and causes Gate 3 to begin discharging. This increased current causes the *virtual ground* to bounce, and consequently, both Gate 1 and Gate 3 slow down. At this point subsequent breakpoints must be updated to reflect the slower circuits. When Gate 1 finishes switching, Gate 3 speeds up, be-

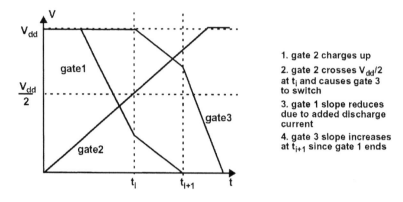

Figure 4.4.   Variable break point switch level simulator function [5]

cause less current needs to be sunk through the sleep transistor. Again, the breakpoints are recomputed at this point to reflect the different operating conditions [5].

In general, the simulator is able to capture the basic effect of sleep transistor sizing on propagation delay, and even though it is based solely on a first order delay model, the simulator still manages to track the switching variations of this MTCMOS circuit. It is evident that the simulator can reasonably track MTCMOS delays as a function of sleep transistor sizings.

Limitations of Switch Level Simulator

The delay model used in the variable breakpoint switch level simulator has the following drawbacks:

- Modeling discharge current as a purely saturation current is incorrect; triode current also exists.

- The impact of parasitic capacitances on the virtual *ground* line is not taken into consideration.

- Complicated gates are modeled as a simple inverter which leads to timing inaccuracies.

- The input patterns that cause large internal node glitching are not modeled.

- Secondary order effects such as velocity saturation, body effect, and parasitic capacitances are not taken into consideration.

4.4. Hierarchical Sizing Based on Mutually Exclusive Discharge Patterns

The hierarchical sizing methodology uses a bottom up approach to synthesize the sleep transistor by ensuring that the performance of every individual gate does not degrade by more than a fixed percentage. By guaranteeing this constraint on individual gates, any combination of gates in a path can also meet or exceed the performance constraint, and thus, the macro performance measure is satisfied.

Example for Hierarchical Sizing [4]

The sleep transistor sizing and merging technique is described by the example in Figure 4.5. An MTCMOS circuit can initially be sized by using individual sleep transistors that can be later merged together into a common, single sleep device for the larger block. The circuit consists of three chains of five low-V_{th} inverters, and measurements are made for the input to output delay, the delay for inverter I_5, and the virtual ground bounce transients.

Individual Sleep Transistor Sizing

Figure 4.5(a) denotes the first step in the transistor sizing procedure; individual resistors (which model sleep transistors in the *on* state) are sized to ensure that no gate degrades by more than a fixed percentage. Columns 2 and 3 of Table 4.1 indicate that the overall performance of the inverter chain will be achieved, if the speed requirements of each individual internal gate are met (which is the case due to properly sizing the resistors); that is, the percentage degradation in delay in Column 3 is always less than or equal to that of Column 2).

Table 4.1. Percentage Degradation for Gate I_5 and Total Delay for Cases (a), (b) and (c), as Function of Sleep Resistance

R ohms	I_5 (a) %	Total (a) %	I_5 (c) %	Total (c) %	I_5 (e) %	Total (e) %
0	0	0	0	0	0	0
100	1.9	1.2	2.0	1.1	4.9	2.9
200	3.0	2.3	3.5	1.9	9.3	5.7
300	4.2	3.4	4.9	2.9	13.7	8.5
400	5.8	4.6	6.1	3.7	17.9	11.0
500	7.5	5.7	7.5	4.6	22.1	13.4

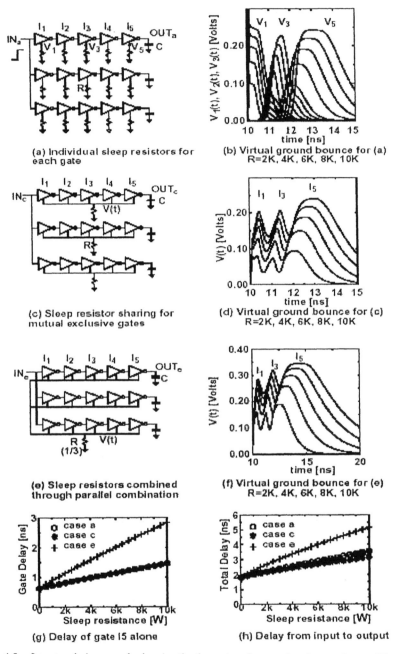

Figure 4.5. Inverter chain example showing the three steps for merging sleep resistors. (Circuit uses V_{dd}=1.0V, V_{th}=0.2V, C=50fF, I_{min}=0.7μm [4])

The overall delay degradation is less than the individual gate delays, because the low to high transitions of I_2 and I_4 are not degraded by the NMOS sleep transistor. Furthermore, the virtual *ground* lines (V_1, V_3, and V_5) for this circuit configuration bounces as a result of a rising step function applied to the input. This is depicted in Figure 4.5(b).

Merging Gates Based on Mutually Exclusive Discharging

Individually adding high-V_{th} transistors to each gate in a circuit would seem to be the simplest method to develop an MTCMOS sizing strategy. However, this can result in not only large overestimation in the sleep transistor area, but also large overheads in the wiring area. Fortunately, not all gates in a circuit switch at the same times. Consequently, it is possible to merge sleep transistors with mutually exclusive gates, and thereby, reduce circuit complexity. For a set of n gates with the equivalent sleep resistances r_1, r_2, ... r_n , the sleep resistors can be grouped and replaced by the single resistance $r_{eff} = $ min (r_1, r_2, ... r_n). Being mutually exclusive, these gates will discharge currents through their common sleep transistor at different times. No overlap in time exists between the discharging currents of the mutually exclusive gates. This causes the virtual ground bounce that each transitioning gate experiences to remain the same or smaller than before. As a result, the delay of each gate sharing the common sleep transistor should also be the same or less than in the original circuit.

Replacing n sleep resistors with a single sleep transistor has several benefits: (1) a reduction of large overheads in the wiring area, (2) a reduction of subthreshold leakage current by a factor of n, and (3) an increased parasitic capacitance on the virtual ground line that reduces noise bounce, and ultimately, improves performance.

Figure 4.5(c) shows how the original inverter tree's sleep resistors can be replaced by only three resistors by utilizing the same high-V_{th} switch for the mutually exclusive gates. For example, inverters I_1, I_2, I_3, I_4, and I_5 will never discharge at the same time, and as a result, can share a common sleep transistor.

It can also be shown that the total path delay in the merged scenario meets or exceeds the performance of the individually sized devices. The slight performance improvement seen in Figure 4.5(h) is due to the larger parasitic capacitance on the virtual ground line after the merge. This tends to low pass filter the virtual *ground* bounce. Therefore, inverter I_1 will be faster in the merged case, because the virtual *ground* bounce rises more slowly due to the added parasitic capacitance. As the parasitic capacitances charges up through, later stage gates will not have these beneficial effects since the capacitance does not have enough time to discharge again.

Merging through Parallel Combination

Separate sleep resistors for different groups of mutually exclusive gates can be cumbersome in the circuit layout. In many cases, it is possible to merge these sleep devices together in a parallel combination with a single switch for the whole circuit, and the performance would still be maintained. To quantify this point, consider the circuit in Figure 4.6.

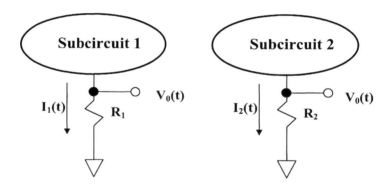

Figure 4.6.   Circuits showing how sleep resistors can be combined in parallel [4]

If the virtual *ground* voltages for two different sub-circuits are similar, then they can be modeled as two current sources, $I_1(t)$ and $I_2(t)$, connected to resistors R_1 and R_2 to produce a voltage waveform $V_0(t)$ on the virtual *ground* rail for both cases. However, if $I_1(t)$ and $I_2(t)$ are summed, and R_1 and R_2 are placed in parallel, then the voltage of the virtual ground rail is

$$\begin{aligned}
V(t) &= [I_1(t) + I_2(t)] \times [R_1//R_2] \\
&= [I_1(t) \times R_1 \times R_2 + I_2(t) \times R_1 \times R_2] \times (1/R_1 + R_2) \\
&= [V_0(t) \times R_2 + V_0(t) \times R_1] \times (1/R_1 + R_2) \\
&= V_0(t)
\end{aligned}$$

which is the same result as before. Therefore, for two sub-circuits with very similar virtual *ground* transient behaviors, combining the two subcircuits results in unchanged virtual ground characteristics, and so the overall circuit performance should be unchanged. In general, if there is a large difference in voltages $v_1(t)$ and $v_2(t)$, then the resistors are combined such that the resultant v(t) does not exceed the minimum of $v_1(t)$ and $v_2(t)$. In this case, the equivalent resistance is

$$R_{eq} = \frac{min[V_1(t), V_2(t)]}{[V_1(t)/R_1 + V_2(t)/R_2]} \tag{4.4}$$

Therefore, the three resistors from Figure 4.5(c) can be combined to form a single resistor with three times the conductance, and now gates the entire circuit as demonstarted in Figure 4.5(e). By scaling the resistance by one third for the case with a single global sleep transistor, the virtual *ground* bounce of Figure 4.5(f) can be matched to that of Figure 4.5(g), which gives the same delay behavior. In general, combining separate sleep transistors into a single one is beneficial; the increased parasitic capacitances will tend to speed up the circuit during the capacitor charging stage.

Random Logic Example

One way to develop a mutually exclusive set of gates in a circuit is to use a criterion based on the structural interconnections in the network graph. If it is assumed that there is a unit delay model for each gate, all the possible times that a gate would switch can be tabulated. Mutually exclusive gates are then grouped together, when there is no overlap between the corresponding sets of times. It is necessary to minimize the number of these groups of mutually exclusive gates to minimize the overall sleep transistor area.

Figure 4.7 illustrates this merge technique for a random logic circuit composed of ten gates with arbitrary gate interconnections. Each individual gate is assumed to have an individual sleep transistor, whose size is adjusted to tolerate a performance degradation of, say 5%. The individual sleep transistor is modeled as a resistor which causes the same 5% degradation in performance. Therefore, ten sleep transistors are employed. Each gate is annotated by using a unit delay model with all the possible time slots where a transition can occur. Gates with no overlap between their time periods are considered mutually exclusive, and can be grouped together with a common sleep transistor. Gates that discharge mutually exclusive are grouped together, and connected to a single sleep transistor whose resistance is the minimum of the individual resistances for each gate in that group. Thus, three sleep transistors are now employed. Finally, the sleep transistors for each group are merged into one

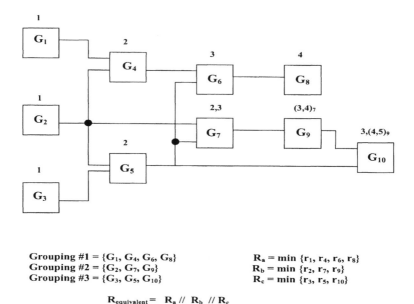

Grouping #1 = {G_1, G_4, G_6, G_8} R_a = min {r_1, r_4, r_6, r_8}
Grouping #2 = {G_2, G_7, G_9} R_b = min {r_2, r_7, r_9}
Grouping #3 = {G_3, G_5, G_{10}} R_c = min {r_3, r_5, r_{10}}

$$R_{equivalent} = R_a \,//\, R_b \,//\, R_c$$

Figure 4.7. Logic gates annotated with all possible transition times, so that sleep resistors can be merged [4]

sleep device whose resistance is computed from the parallel combination of the existent sleep transistor. The result is a single sleep transistor that supports the whole circuit.

This design methodology which is based on mutually exclusive discharge patterns is most effective for balanced circuits with minimal glitching. This is possible, because balanced circuits are composed of many gates that discharge exclusively. An example of such circuits is the 32-bit parity checker circuit in Figure 4.8. With the application of the merge algorithm, the total number of sleep transistor is reduced from 31 (one for each gate) to 16.

4.5.   Designing High-V_{th} Sleep Transistors, the Average Current Method [6]

Another technique to estimate the optimum size of the sleep transistor is based on an *average current method*. If the average current flowing through the sleep transistor, and a maximum speed penalty for the MTCMOS block are known, then a minimum size for the sleep transistor can be estimated. The

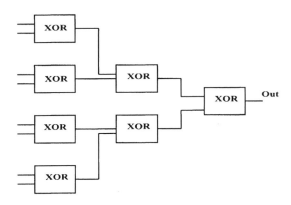

Figure 4.8. 8-bit parity checker

technique relies on the fact that the current consumed in the MTCMOS circuit is constant with the circuit's operating pattern, and hence, the ΔV is also approximately constant. ΔV is the voltage drop across the sleep transistor (Figure 4.9(a)), which consequently implies that the MTCMOS speed penalty does not change either.

In order to justify that the MTCMOS speed penalty does not strongly depend on the circuit operation patterns if the speed penalty is set low, the following experiment is conducted. Figure 4.10(a) represents the simulation results for various operation patterns, A, B, and C, that are shown in Figure 4.9(c).

In Pattern A, a large number of gates are switching within a short period of time. Pattern C has a small number of switching gates in the clock period, and Pattern B is in-between the two patterns. For a small size sleep transistor, the delay penalty of Pattern A is larger than that of Pattern C. This is explained by Figure 4.10(b) which presents transient simulations of the virtual *ground* rail for each of the three patterns for an MTCMOS circuit, operating at a supply voltage of 1V. Since the sleep transistor is sized small (W_{sleep}=0.2mm for a 0.25μm MTCMOS/SIMOX technology), the simultaneous gate switching in Pattern A causes the virtual *ground* rail to bounce as much as 0.5V (half V_{dd}). Furthermore, the virtual *ground* bounce spans a long period of time. When the sleep transistor is sized up to W_{sleep}=2mm, the delay penalty of Pattern A is almost the same as that of Pattern C. The large sleep transistor acts as a current reservoir that suppresses large *ground* bounces, and will not sweep a long time period.

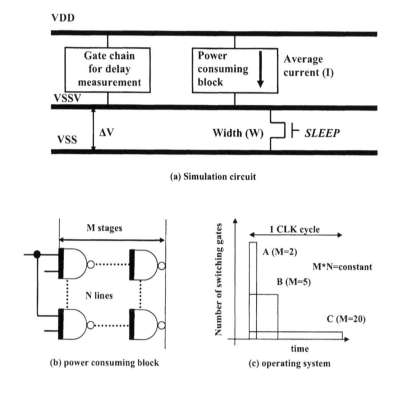

Figure 4.9.   Simulation for MTCMOS delay [6]

Therefore, it can be concluded that for sufficient sleep transistor sizes (the low speed penalty), the speed penalty does not depend on the circuit operating pattern. This makes it easy to analyze the speed penalty, since the voltage drop at the sleep transistor and gate time degradation are now treated as static phenomena.

W_{sleep} **Estimation**

It is now necessary to estimate the sleep transistor size by knowing the average current flowing through it. The gate delay time of a CMOS circuit at a supply voltage of V_{dd} is, approximately, expressed as

$$\tau(V_{dd}) \propto \frac{CV_{dd}}{\beta(V_{dd} - V_{tL})^{\alpha}} \qquad (4.5)$$

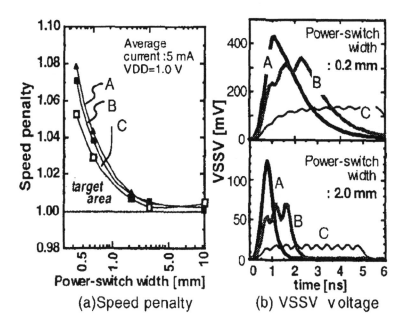

Figure 4.10.   Simulation results presented in [6]

where β is the drivability factor, C is the output load capacitance, α is the saturation index, and V_{tL} is the low-V_{th}.

As previously illustrated, the effective supply voltage of the MTCMOS circuit is V_{dd}-ΔV. It is assumed that the low-V_{th} CMOS circuit with a V_{dd}-ΔV supply has the same gate delay time as an MTCMOS circuit. The MTCMOS Speed Penalty (MSP) expresses the ratio of the delay times at V_{dd}-ΔV and V_{dd} power supplies as

$$MSP = \frac{\tau[V_{dd} - \Delta V]}{\tau[V_{dd}]} \approx [\frac{V_{dd} - V_{tL}}{V_{dd} - V_{tL} - \Delta V}]^{\alpha-1} \qquad (4.6)$$

ΔV is basically $R_{sleep} \times I_{sleep}$, where R_{sleep} and I_{sleep} are the sleep transistor resistance and the average current flowing through the sleep transistor, respectively. Moreover, R' is the normalized sleep transistor resistance, and the actual resistance R is determined by $R=R'/W_{sleep}$. The MSP in Equation (4.6) can be rewritten as

$$MSP = \frac{1}{[1 - \frac{I_{sleep}}{W_{sleep}} - \frac{R'}{V_{dd}-V_{tl}}]^{\alpha-1}} \qquad (4.7)$$

Therefore, from Equation (4.7), the size of the sleep transistor is determined by knowing the average current flowing through it,

$$W_{sleep} = \frac{1}{1 - MSP^{\frac{1}{1-\alpha}}}[\frac{R'}{V_{dd} - V_{tL}}]I_{sleep} \qquad (4.8)$$

Equation (4.8) provides an estimate of the sleep transistor size, because it contains several approximations. In order to determine a more accurate sleep transistor size, two issues should be determined: (1) the delay penalty dependency on ΔV, and (2) the resistance value of the sleep transistor R_{sleep} which is determined from the I-V characteristics of the sleep transistor.

If a 2% delay penalty is allowed, then ΔV is 2% of V_{dd}. For a 1V power supply, ΔV is equal to 20mV. However, R_{sleep} is evaluated by using the slope of the I-V line of the sleep transistor, shown in Figure 4.11. With a sleep transistor width W_{sleep} of 1mm, R_{sleep} is approximately 3Ω, which leads a R' value of 3Ωmm.

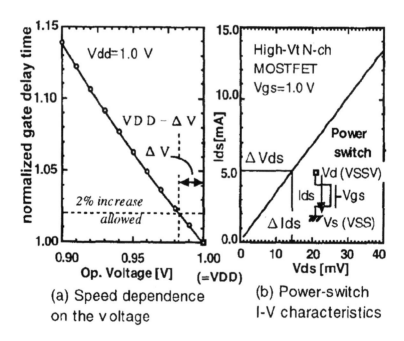

Figure 4.11. Basic data for the ACM presented in [6]

The required switch size can now be determined by

$$W_{sleep} = \frac{R'}{\Delta V} \times I_{sleep}$$
$$= \frac{3\Omega mm}{20mV} \times I_{sleep}(mA)$$
$$= 0.15 \times I_{sleep}$$

A 1V, 290K gate LSI designed in a 0.25μm MTCMOS process [6], to verify the effectiveness of this sizing technique. The conventional sizing method would require a sleep transistor with a width of 1000mm to contain the 290K gate LSI. However, by using the average current method illustrated in Figure 4.11, the sleep transistor has a width of only 160mm. This size reduces the standby leakage current from 6μA to 1μA.

4.6. Drawbacks of Techniques

In [4], cascaded gates are clustered together, because simultaneous current discharges can never take place. This methodology may be efficient for balanced circuits with tree configurations, where mutually exclusive discharging gates are easily detected. However, this methodology would not be as efficient for circuits with complicated interconnections and unbalanced structures. Therefore, sleep transistor assignments can therefore be wasteful, and cause both the dynamic and leakage power to rise. Finally, the sets of sleep transistors in [4] are merged into a single large sleep transistor to accommodate the entire circuit as in [7]. In addition to these drawbacks, a single sleep transistor containing the whole circuit increases the interconnect resistance for the distant blocks. As a result, the sleep transistor must be sized even larger than expected to compensate for the added interconnect resistance. Excessively large sleep transistors again augment dynamic and leakage power, as well as the area. This drawback would be even more severe in DSM regimes, where interconnects have a large impact on the circuit's performance [8]. The authors' proposed methodologies are presented in the upcoming sections to solve this problem. Not only gates with exclusive discharge patterns are considered, but also with partially overlapping discharge currents.

4.7. Distributed Sleep Transistors [9] [10]
Sizing the Sleep Transistor

The first step in the proposed techniques is to estimate the size of the sleep transistor. To do this, the delay of a single gate τ_d without a sleep transistor can be expressed as [1] [7]

$$\tau_d \propto \frac{C_L V_{dd}}{(V_{dd} - V_{tL})^\alpha} \qquad (4.9)$$

where C_L is the load capacitance at the gate's output, V_{tL} is the low-V_{th} (LVT)=350mV, V_{dd}=1.8V, and α is the velocity saturation index which is equal to ≈ 1.3 in 0.18μm CMOS technology. In the presence of a sleep transistor, the delay of a single gate τ_d^{sleep} can be expressed as

$$\tau_d^{sleep} \propto \frac{C_L V_{dd}}{(V_{dd} - V_X - V_{tL})^\alpha} \qquad (4.10)$$

where V_X is the potential of the virtual *ground* shown in Figure 4.1. If it is assumed that the circuit can tolerate a 5% degradation in performance due to the presence of the sleep transistor, then

$$\frac{\tau_d}{\tau_d^{sleep}} = 95\% \qquad (4.11)$$

If Equations (4.9) and (4.10) are substituted in Equation (4.11), and taking $\alpha = 1$ for simplicity [1], then

$$1 - \frac{V_X}{(V_{dd} - V_{tL})} = 95\% \qquad (4.12)$$

Therefore, V_X can be formulated as

$$V_X = 0.05(V_{dd} - V_{tL}) \qquad (4.13)$$

The current flowing through the $linearly - operating$ sleep transistor is expressed as

$$\begin{aligned} I_{sleep} &= \mu_n C_{ox}(W/L)_{sleep}[(V_{dd} - V_{tH})V_X - V_X^2/2] \\ &\approx 0.05\mu_n C_{ox}(W/L)_{sleep}(V_{dd} - V_{tL})(V_{dd} - V_{tH}) \end{aligned} \qquad (4.14)$$

where μ_n is the N-mobility, C_{ox} is the oxide capacitance, and V_{tH} is the HVT=500mV. The size of the sleep transistor can now be written as

$$(W/L)_{sleep} = \frac{I_{sleep}}{0.05\mu_n C_{ox}(V_{dd} - V_{tL})(V_{dd} - V_{tH})} \qquad (4.15)$$

The values for I_{sleep}, and consequently $(W/L)_{sleep}$, are chosen to exhibit low-power dissipation. Thus, an optimization problem exists to find the value for

I_{sleep}, and consequently $(W/L)_{sleep}$, to dissipate minimal dynamic and leakage power. This will be detailed in Section 4.8. To illustrate the basic idea behind the proposed techniques, a value for I_{sleep} is chosen to be $250\mu A$, leading to a $(W/L)_{sleep} \approx 6$ (Equation(4.15)) for a $0.18\mu m$ CMOS technology. This constant size $(W/L)_{sleep} = 6$ will be first used for illustrating the first two proposed methodologies, that is, Bin-Packing (BP) and Set-Partitioning (SP) techniques. To ensure the correct functionality, delay, power and leakage values which a circuit has, analytical calculations are verified for the LVT and HVT HSPICE models. Leakage current increases by an order of magnitude for every 85mV reduction in V_{th}.

4.8. Clustering Techniques

To illustrate the clustering techniques, six benchmarks are used as test vehicles; a 4-bit Carry Look Ahead (CLA) adder, a 32-bit priority checker, a 6-bit array multiplier design, a 4-bit ALU/Function Generator (74181 ISCAS-85 benchmark), a 32-Single Error Correcting circuit (C499 ISCAS-85 benchmark), and a 27-bit Channel Interrupt Controller (CIC) (C432 ISCAS-85 benchmark). These benchmarks are chosen to offer a variety of circuits with different structures that employ various gates with different fanouts. The 4-bit CLA adder will be the first benchmark used to demonstrate the proposed techniques. The results pertaining to all the other benchmarks will be provided at the end of this section.

Figure 4.12 presents a schematic diagram of the CLA adder which consists of 28 gates (G_1-G_{28}). All the gates are implemented in $0.18\mu m$ CMOS technology. To illustrate the proposed technique, a preprocessing stage of gate currents is described in the next section. The preprocessing stage will be utilized to solve the BP problem, and later on to solve the SP problem.

Preprocessing of Gate Currents

The main objective of the preprocessing stage is to group the gates into subclusters such that their combination does not exceed the maximum current of any gate within the cluster.

All the gates used in the implementation of the benchmarks are based on the $0.18\mu m$ standard cell library by *Virtual Silicon Technology Inc.* which uses the $0.18\mu m$ $TSMC$ Process. In this library, 250 logic gates are defined. For each logic gate, different propagation delays are documented according to: (1) the input vector applied to the gate, (2) the kind of transition occurring at the input of the gate, (3) the amount of fanout associated with the gate, and (4) whether

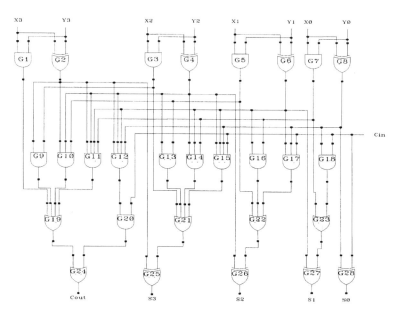

Figure 4.12.   4-bit Carry Look-Ahead adder

the signal at the output of the gate is falling or rising. In summary, the standard cell library carries all the gate information. The documented propagation delays for the standard cell library are verified through HSPICE simulations under the same environment (i.e., the same 0.18μm CMOS technology provided by $TSMC$; V_{dd}=1.8V and temperature=25°C).

In the analysis of our proposed techniques, only the propagation delay for an output falling edge signal is taken into consideration, because the analysis takes into account the current flowing through the sleep transistor.

Each tested benchmark is composed of gates with different fanouts. Each of the gates is mapped to the documented Standard Cell Library according to its functionality and fanout. For each gate, all the input combinations are applied e.g., 00,01,10,11 for a 2-input XOR gate), and the highest discharging current at the output of every gate is monitored (the worst case discharge current), while taking the gate's fanout into consideration. The discharge current, as well as the short-circuit currents, are monitored, because this is the current that flows through the sleep transistor and eventually the *ground*. The probability that discharging takes place (the switching activity) at the output of each gate is calculated and multiplied by the corresponding discharge peak current. This gives an *expected* discharge current value. The switching activity of a

gate is computed by multiplying the probability that the output of the gate will
be at "0", by the probability that the output will be at "1" [11]. If the switching
activity is not accounted for, the design problem will be very pessimistic and
the sleep transistor will be oversized, causing substantial increase in leakage
and dynamic power dissipation, as well as, in the die size. It is very unlikely
that the clustered gates will have their worst case current discharge at the same
time. This has been deduced by exhaustively applying all the input vectors
to the CLA adder benchmark. The monitored current is composed of the discharge and the short-circuit currents that take place during switching. Sleep
transistors should be sized to also accommodate the short-circuit currents; otherwise, the speed will be degraded.

The peak current value and time at which the switching occurs, as well as
its duration, are monitored. The time that the switching takes place depends
on the gate's propagation delay and input pattern, whereas the current duration
depends on the slope of the input signal, as well as the fanout of the gate. The
larger the input slope and/or gate fanout, the longer the switching duration.
The discharge current of each gate takes a triangular shape, whose peak occurs at a time equal to the gate delay, and spans a time, mainly function in the
fanout of the gate. Since the switching activity of a gate is a constant number,
multiplying it by the triangular shaped discharge current would also produce
a triangular shape spanning the same time duration, but with a smaller peak
value.

To facilitate the vector comparisons and to offer an automated design environment, every discharge current at the output of a gate is represented by a
vector. The time axis is divided into time slots, each equal to 10psec, as represented in Figure 4.13. A time slot of 10psec is sufficient in 0.18μm CMOS
technology to offer a relatively good accuracy. Each time slot holds a value
that represents the magnitude of the discharge current at that specific time that
constitutes an element in the vector. Figure 4.13 portrays this idea, with a 2-
input AND gate (G_1) with a fanout of two driving a 2-input OR gate (G_2) with
a fanout of four. Furthermore, a load of 6fF is applied to the outputs of each
circuit contributing to the wiring capacitance. The discharge currents of G_1
and G_2 (I_1 and I_2) are each presented as a vector.

Each element in the vector presents the magnitude of current at this 10psec
time slot. The peak of the discharge current for G_1 occurs at the gate's delay
time ($T_1 = 80psec$), while the discharge current I_2 occurs at time ($T_1 + T_2 = 210psec$), because G_2 will not discharge until G_1 discharges. The peak cur-

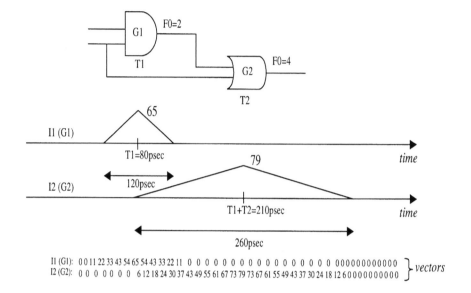

II (GI): 0 0 11 22 33 43 54 65 54 43 33 22 11 0 0 0 0 0 0 0 0 0 0 0 0 0 0 0 0 000000000000 } vectors
I2 (G2): 0 0 0 0 0 0 0 6 12 18 24 30 37 43 49 55 61 67 73 79 73 67 61 55 49 43 37 30 24 18 12 6 0 0 0 0 0 0 0 0 0 0

Figure 4.13.   Timing diagram

rents of gates G_1 and G_2 are $65\mu A$ and $79\mu A$, respectively. The triangular shaped currents are converted into vectors as indicated in Figure 4.13. Since G_2 has a large fanout of four, the duration of the discharge current is long (260 psec), while the duration of the discharge current in G_1 is short due to the small fanout of two (120 psec). Therefore, for every gate in the circuit, a vector is constructed that carries information about the delay of the gate (when the peak occurs), the fanout of the gate (the duration of the current), and the magnitude of the current in each time slot. By constructing a vector for each gate, a series of vectors (28 in this case) that carry information about the whole circuit are produced.

Trapezoidal Current Discharge

In Figure 4.13, the current I_1 is assumed to be discharged at a time equal to the average delay T_1 of G_1 (provided that G_1 is a primary output gate in the circuit). Similarly, current I_2 is assumed to be discharged at time $T_1 + T_2$, where T_2 is the gate G_2 average delay. However, the delay of any gate, and consequently the delay of the whole circuit, changes with the input vectors. For example, consider a 2-input NAND gate in Figure 4.14. The high-to-low propagation delay (affected by the sleep transistor) varies with the different changes in the input vector[2].

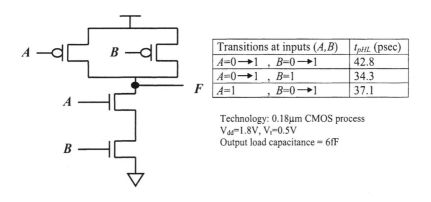

Transitions at inputs (A,B)		t_{pHL} (psec)
$A=0\rightarrow1$, $B=0\rightarrow1$		42.8
$A=0\rightarrow1$, $B=1$		34.3
$A=1$, $B=0\rightarrow1$		37.1

Technology: 0.18μm CMOS process
V_{dd}=1.8V, V_t=0.5V
Output load capacitance = 6fF

Figure 4.14. 2-input NAND example

Consequently, a method should be provided that insures that the current discharge is taken into account with any input vector combination. Accounting for the discharge current over all the input vector combinations guarantees that the sleep transistor is sized properly, and that the circuit meets the target performance. Therefore, the $min/max\ technique$ is applied. The objective of this technique is to determine the earliest and latest times that a current discharge takes place at the output of a certain gate. It is seen that the 2-input NAND example in Figure 4.14, t_{min}=34.3 psec (the earliest time), while t_{max}=42.8psec (the latest time), where t_{min} and t_{max} are the minimum and maximum high-to-low propagation delays of the NAND gate. The min/max technique is further explained by the following example:

Consider the random logic circuit in Figure 4.15. A parameter T_k is defined, which denotes the accumulative delay at the output of gate G_k. The accumulative delay of gate G_k has a minimum value $T_{k_{min}}$ and a maximum value $T_{k_{max}}$, whose values depend on the accumulative delays of the preceding gates. For a primary output gate G_k such as G_1, G_2, G_3 and G_4, $T_{k_{min}}$ is equal to $t_{k_{min}}$, where $t_{k_{min}}$ is the minimum intrinsic gate delay of G_k. On the other hand, $T_{k_{max}}$ is equal to $t_{k_{max}}$, where $t_{k_{max}}$ is the maximum intrinsic gate delay of G_k.

For example, for a primary output gate such as G_2, $T_{2_{min}}=t_{2_{min}}$, while $T_{2_{max}}=t_{2_{max}}$. If G_2 is a 2-input NAND gate, $T_{2_{min}}=t_{2_{min}}=34.3$psec, and

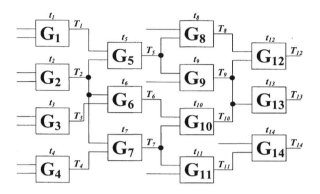

Figure 4.15. Random logic network example

$T_{2_{max}} = t_{2_{max}} = 42.8$psec (Figure 4.14). For non-primary output gates, $T_{k_{min}}$ and $T_{k_{max}}$ the minimum and maximum delays for each input to that non-primary output gate need to be considered . For example, gate G_6 is a non-primary output gate fed by gates G_2 and G_3. Therefore, $T_{6_{min}}$ and $T_{6_{max}}$ can be written as

$$T_{6_{min}} = min\{(T_{2_{min}} + t_{6_{min}}), (T_{3_{min}} + t_{6_{min}})\} \qquad (4.16)$$

and

$$T_{6_{max}} = max\{(T_{2_{max}} + t_{6_{max}}), (T_{3_{max}} + t_{6_{max}})\} \qquad (4.17)$$

Similarly, the accumulative delay T_{10} of the non-primary output gate G_{10} is

$$T_{10_{min}} = min\{(T_{6_{min}} + t_{10_{min}}), (T_{7_{min}} + t_{10_{min}})\} \qquad (4.18)$$

and

$$T_{10_{max}} = max\{(T_{6_{max}} + t_{10_{max}}), (T_{7_{max}} + t_{10_{max}})\} \qquad (4.19)$$

In general, the accumulative delay T_k of gate G_k is expressed as

$$T_{k_{min}} = min\{(T_{i_{min}} + t_{k_{min}}), (T_{j_{min}} + t_{k_{min}}), ...\} \qquad (4.20)$$

and

$$T_{k_{max}} = max\{(T_{i_{max}} + t_{k_{max}}), (T_{j_{max}} + t_{k_{max}}), ...\} \qquad (4.21)$$

given that the output of gates G_i, G_j, ... are the inputs to G_k. Equations (4.20) and (4.21) can be rewritten as follows:

$$T_{k_{min}} = min\{(T_{i_{min}} + t_{k_{min}})\} \forall i = 1, 2, 3, ...n \qquad (4.22)$$

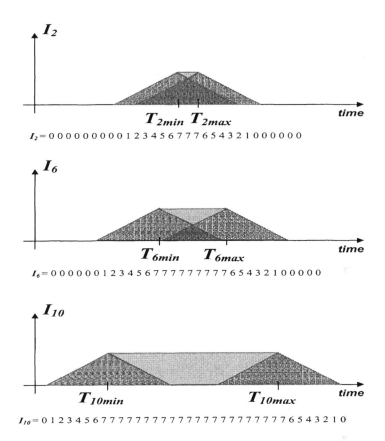

Figure 4.16.   Discharge currents in the random logic example

and

$$T_{k_{max}} = max\{(T_{i_{max}} + t_{k_{max}})\} \forall i = 1, 2, 3, ...n \qquad (4.23)$$

where n is the number of inputs to gate k.

The general expression for the accumulative min and max delays $T_{k_{min}}$ and $T_{k_{max}}$ in Equations (4.20) and (4.21) are also valid for the special case that gate G_k is a primary output gate. In that case, $T_{i_{min}}$, $T_{j_{min}}$, $T_{i_{max}}$, and $T_{j_{max}}$ are equal to $zero$, leading to $T_{k_{min}} = t_{k_{min}}$ and $T_{k_{max}} = t_{k_{max}}$.

In Figure 4.13, the peak discharge current is assumed to occur at a single time value. Since the delay of a gate changes with the input vector, the discharge current must be taken into account during the time period from $T_{k_{min}}$ to $T_{k_{max}}$ for each gate G_k. This guarantees that regardless of the input vector, the discharge current is taken into account, and the speed of the circuit is attained. To illustrate this, Figure 4.16 shows a rough diagram of the discharge currents for gates G_2, G_6, and G_{10} in Figure 4.15. Figure 4.16 is for illustration purposes only, and does not take into account the actual discharge current values or how fanout changes the duration of the discharge. The triangular shaped discharge current, previously shown in Figure 4.13, sweeps the time range from $T_{k_{min}}$ to $T_{k_{max}}$. Therefore, the discharge current is no longer modeled as a triangle, but as a trapezoid which accounts for the variation in delay due to changes in the input vector. The corresponding vectors for currents I_2, I_6 and I_{10} are also shown in Figure 4.16. It is interesting to note from this figure that the first stage gates (the primary output gates) G_1, G_2, G_3 and G_4 would have discharge currents that sweep short time durations. Deeper in the circuit, the discharge current would span longer time durations for the gates located in later stages. This is attributed to the increase in the accumulated delay in the gates located in the later stages.

This method insures that the current discharge is taken into account for any input vector combination. As explained, accounting for the discharge current over all the input vector combinations guarantees that the sleep transistor will be sized properly, and that the circuit will meet the target performance. Although the approach presented may seem to pessimistically model the discharge currents, which may increase the size or number of sleep transistors, it actually proves useful in accounting for the following two important factors:

- **Intrinsic Gate and Interconnect Delays**: The delay of a circuit varies with the variation in the intrinsic gate delay, as well as the interconnect delay. These variations are more severe in the deep-submicron regime. Intrinsic gate delays vary due to variations such as threshold-voltages, transistor dimensions, doping concentrations, and input signal slope variations [12]. Furthermore, interconnect delays vary due to variations in wire dimensions and coupling noise [13]. The discharge current from $T_{k_{min}}$ to $T_{k_{max}}$, as illustrated in Figure 4.16, will include all discharge currents even if the circuit delay varies. This guarantees that the performance is not degraded with any delay variations.

- *Glitching Currents*: Glitching currents usually arise at the output of a gate whose inputs do not arrive at the same time. In Figure 4.13, glitching currents may arise at the output of gate G_2 due to the unbalanced delay paths for the inputs. If these glitching currents are not taken into account when the sleep transistor is sized, performance will be affected at the time that these glitching currents occur. In min/max technique, the discharge current is accounted for, from the minimum delay of all the inputs $T_{k_{min}}$ to the maximum delay of all the inputs $T_{k_{max}}$. This insures that any glitching currents are taken into consideration, and thus the sleep transistors are sized properly to fulfill the target performance.

Preprocessing heuristic

Figure 4.17 denotes the preprocessing heuristic that forms a set of subclusters of gates that will not exceed the maximum current of any gate within the cluster when they are combined [9].

```
              PREPROCESSING HEURISTIC
1. Initialize current vectors
2. Set all Gates free; to move to subcluster;
3. For all gates in circuit
        If gate G_i is not clustered yet
            assign gate G_i to new cluster C_k
            update cluster current vector
            calculate max current,start,end time
        End If
        For all other gates in circuit
            If (gate G_j is not clustered yet)
                add current of gate G_j to cluster C_k
                If (combination ≤ max current)
                    append gate to cluster
                    update cluster info
                    set gate G_j locked in cluster C_k
                End If
            End For
        End For
4. Return all clusters formed.
```

Figure 4.17.   Heuristic for Forming Sub-clusters

First, the preprocessing algorithm initializes the current vectors of all the gates as described in Figure 4.16. At the beginning of the preprocessing algorithm all the gates are set free to move to any newly created cluster. Once a gate is collapsed into a cluster, the gate is locked in, and unable to participate in the formation of new clusters. Then, the cluster information (in terms of the number of gates, current, and maximum current) is updated. The criteria used to append a gate to a cluster is based on the maximum current capacity of the current cluster. As shown in step (3) of the algorithm found in Figure 4.16, a cluster is initially seeded with a free gate after which all the other gates are appended to the cluster.

The preprocessing algorithm terminates when it is not possible to append any more gates to a cluster. Table 4.2 lists the results of applying the preprocessing heuristic to the 4-bit CLA adder that was presented in Figure 4.12, where 14 sub-clusters are formed. For example, the second column in Table 4.2 (I_{EQ_2}) represents a subcluster, formed by combining Gates G_3, G_4, G_{27} and G_{28} which has a maximum current of $80\mu A$ of the partially overlapped discharging gates ($I_{G3}^{max} = 65\mu A$, $I_{G4}^{max} = 80\mu A$, $I_{G27}^{max} = 34\mu A$ and $I_{G28}^{max} = 30\mu A$). $I_{overlap}^2 = max\{I_{G3}^{max}, I_{G4}^{max}, I_{G27}^{max}, I_{G28}^{max}\} = I_{G4}^{max} = 80\mu A$. The objective is then to group as much current (as many gates) as possible without exceeding the current limit of the sleep transistor ($250\mu A$), while the number of sleep transistors used is minimized. This is shown in Table 4.3. This problem presentation is analogous to the Bin-Packing problem in operations research which will be discussed next.

Table 4.2.   Results: Current Equivalence

I_{EQ_1} $I_{overlap}(\mu A)=80$	I_{EQ_2} 80	I_{EQ_3} 110	I_{EQ_4} 90	I_{EQ_5} 50	I_{EQ_6} 30	I_{EQ_7} 16
$I_1, I_2, I_{12}, I_{19}, I_{20},$ $I_{21},\ I_{22},\ I_{24},\ I_{25},$ I_{26}	$I_3,$ $I_4,$ $I_{27},$ I_{28}	$I_5,$ I_6	$I_7,$ I_8	I_9	I_{10}	I_{11}
I_{EQ_8} $I_{overlap}(\mu A)=50$	I_{EQ_9} 30	$I_{EQ_{10}}$ 16	$I_{EQ_{11}}$ 50	$I_{EQ_{12}}$ 30	$I_{EQ_{13}}$ 50	$I_{EQ_{14}}$ 37
I_{13}	I_{14}	I_{15}	I_{16}	I_{17}	I_{18}	I_{23}

The Bin-Packing Technique

The Bin-Packing (BP) problem [14] can be described as follows: given n *items* (a set of equivalent *currents* in this case) and m *bins* (*sleep transistors* in this case), with

$$I_{EQ_j} = \text{equivalent } current \text{ of gate } j$$
$$\text{and} \quad I_{max} = capacity \text{ of each } sleep\ transistor = 250\mu A$$

The objective is to assign each I_{EQ} to one bin so that the total current in each bin does not exceed I_{max}, and the number of bins used is minimized. It is important to notice that the peak current of a combination of logic gates *subcluster* which will be described in Table 4.3, is directly related to the peak current of the individual logic gates.

The mathematical formulation of the problem is as follows:

$$Minimize\ z = \sum_{i=1}^{m} y_i \qquad (4.24)$$

subject to

$$\sum_{j=1}^{n} I_{EQ_j} x_{ij} \leq I_{max} y_i, \quad i \in \{1, ..., m\},$$

$$\sum_{i=1}^{m} x_{ij} = 1, \qquad (4.25)$$

where

$$y_i = \begin{cases} 1, & \text{if bin } i \text{ is used} \\ 0, & \text{otherwise} \end{cases} \qquad x_{ij} = \begin{cases} 1, & \text{if items } j \in \text{bin } i; \\ 0, & \text{otherwise} \end{cases}$$

This model is a pure Binary Integer Programming problem (BIP). The objective function to be minimized, z, is analogous to the minimum number of sleep transistors used. y_i is analogous to the sleep transistors available. x_{ij} takes a value of "1" if current I_{EQ_j} is assigned to bin i. CPLEX 7.5, a commercial ILP solver, is used to solve this BP problem to determine which currents should be grouped together, and to which sleep transistor they are assigned. The current assignments are summarized in Table 4.3.

It is clear from Table 4.3 that three sleep transistors will be needed to contain

Table 4.3. Results: Current Assignments

Sleep Transistor (Cluster)	1	2	3
Equivalent Currents	$I_{EQ_5}, I_{EQ_7}, I_{EQ_8}, I_{EQ_9}, I_{EQ_{10}}, I_{EQ_{11}}, I_{EQ_{12}}$	$I_{EQ_1}, \quad I_{EQ_6}, I_{EQ_{14}}$	$I_{EQ_3}, \quad I_{EQ_4}, I_{EQ_{13}}$
Assigned Gates	$G_9, G_{11}, G_{13}, G_{14}, G_{15}, G_{16}, G_{17}$	$G_1, G_2, G_3, G_4, G_{10}, G_{12}, G_{19}, G_{20}, G_{21}, G_{22}, G_{23}, G_{24}, G_{25}, G_{26}, G_{27}, G_{28}$	$G_5, G_6, G_7, G_8, G_{18}$
$\sum$ Currents(μA)	242	227	250

all the gates in the circuit ($z = 3$). It should be noted that the total current of any cluster must never exceed the maximum current limit of the sleep transistor which is 250μA.

Now that the basic idea for the BP technique has been exemplified through I_{sleep}=250μA, the optimum I_{sleep} value which dissipates the least dynamic and leakage power must be found. Six values for I_{sleep} are considered: I_{sleep}= 150, 200, 250, 300, 350, 400μA. Table 4.4 charts the different values for I_{sleep}, and the corresponding $(W/L)_{sleep}$ and W_{sleep}, calculated by Equation(4.15), where L_{sleep} is taken as 180nm for a 0.18μm CMOS technology.

Table 4.4.   Values for I_{sleep}

I_{sleep} (μA)	150	200	250	300	350	400
$(W/L)_{sleep}$	3.67	4.89	6	7.5	8.56	9.78
W_{sleep} (μm)	0.66	0.88	1.1	1.32	1.54	1.76

The six benchmarks that are chosen are simulated with different sleep transistor sizes listed in Table 4.4. In each case, both the dynamic and leakage power ($P_{dynamic}$ and $P_{leakage}$) are calculated, and a Figure Of Merit (FOM) is generated which is the product of $P_{dynamic}$ and $P_{leakage}$.

The dynamic power is calculated during the active mode ($SLEEP$ signal controlling sleep transistor="1" in Figure 4.1, that is, the sleep transistor is *on*). The current drawn from the supply is monitored for all the generated input vectors and averaged, and then multiplied by V_{dd} to produce the average $P_{dynamic}$. The dynamic power dissipated due to the *on* and *off* switching of the sleep transistors is ignored, since the standby/sleep time period is significantly longer than the active period [15]. On the other hand, the leakage power is calculated during the standby mode ($SLEEP$ signal="0"; the sleep transistor is *off*). All the input vector combinations are applied to the circuit inputs, and the measured leakage current is monitored for each input vector, and then averaged. The average leakage current is then multiplied by V_{dd} to get the average $P_{leakage}$. The leakage power is averaged because $P_{leakage}$ varies with the input vector [16].

This FOM is plotted for the different sleep transistor sizes for all six benchmarks. The sleep transistor size which achieves the minimum FOM is recorded. Figure 4.18 plots the normalized FOM ($P_{dynamic} \times P_{leakage}$) versus I_{sleep} (W_{sleep}) for the 6-bit multiplier. The FOM curve is normalized to its minimum value over different I_{sleep} values. A W_{sleep} of approximately 1.32μm ($I_{sleep} \approx 300\mu$A) achieves the minimum FOM for the 6-bit multiplier case. Furthermore, Figure 4.18 plots the normalized $P_{leakage}$ over different sleep transistor sizes. The $P_{leakage}$ curve is normalized to its minimum value over different I_{sleep} values. The minimum $P_{leakage}$ is achieved with the same sleep transistor size ($W_{sleep} \approx 1.32\mu$m) as the FOM, because dynamic power is not affected to the first order by the choice of sleep transistor size.

Since at a sleep transistor size $W_{sleep} = 1.32\mu$m, $P_{leakage}$ and FOM are minimized, the values of $P_{dynamic}$ and $P_{leakage}$ at $W_{sleep} = 1.32\mu$m are recorded. This cycle is repeated for the remaining five benchmarks. From Figure 4.18, it can be seen that when W_{sleep} (I_{sleep}) takes a very small value, the number of sleep transistors (ST) increases (ST=6 for $W_{sleep}=0.66\mu$m), and consequently, both $P_{dynamic}$ and $P_{leakage}$ are augmented. As I_{sleep} takes higher values, more gates can be confined within a sleep transistor. Even though the sleep transistor size increases, the large reduction in the number of sleep transistors, causes $P_{dynamic}$ and $P_{leakage}$ to drop. Eventually, a stage is reached that as I_{sleep} increases, the savings in the ST number is reduced in proportion to the increase in W_{sleep}. This will cause $P_{dynamic}$ and $P_{leakage}$ to augment again. Thus, the curve in Figure 4.18 has a point where the $P_{dynamic}$, $P_{leakage}$ product has a minimum value. In addition, the minimum $P_{leakage}$ is achieved at that same point. This occurs because $P_{leakage}$ is directly proportional to W_{sleep}, unlike $P_{dynamic}$. It should also be noted that based on the minimum FOM value achieved for each benchmark, W_{sleep} (I_{sleep}) varies from one benchmark to another, depending on the benchmark's structure and circuit topology. For example, the 4-bit ALU benchmark has a minimum FOM at $I_{sleep}=200\mu$A ($W_{sleep}=0.88\mu$m). This is different than the 6-bit multiplier case at hand, where the minimum FOM is achieved at $I_{sleep}=300\mu$A ($W_{sleep}=1.32\mu$m). The values for W_{sleep} to achieve a minimum FOM for the benchmarks are summarized in Table 4.5 later in this section.

With the 5% speed degradation as the comparison basis, the BP technique is compared to [7] and [4]. The operational frequency is 500MHz, and a load of 6fF is applied to the outputs of each gate in a benchmark.

The results are mentioned later in this section and are summarized in Table 4.5 (normalized to [7]). The BP technique proves to attain a high dynamic and leakage power savings, an average 85% and 95% leakage savings compared to that of [4] and [7], respectively. In addition to the reduction in leakage power, the BP method reduces the dynamic power by an average of 17% with respect to [7], and 14% with respect to [4]. The leakage power is calculated in the standby mode, when the sleep transistors are off ($SLEEP$="0") and the

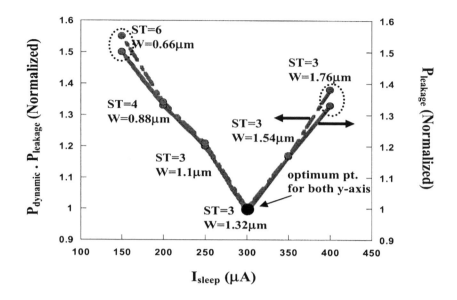

Figure 4.18. Different I_{sleep} values for the 6-bit Multiplier

inputs are inactive.

The BP technique is particularly efficient when it is applied to small circuits that have unbalanced structures. One limitation is that the BP technique does not take the physical locations of the gates on the chip into consideration. For larger circuits, this may cause two gates located far apart to be clustered together which will augment the routing complexity of the circuit. The SP technique solves this problem, and consequently, reduces the routing complexity of the circuit, unlike [7] and [4].

The Set-Partitioning Technique

The Set-Partitioning (SP) problem [14] can be described as follows. Similar to the BP problem, m currents (gates) are arranged into groups such that each gate is included only once in a cluster. A cost function c_j is associated with each group j, S_j. The cost function c_j is evaluated from the physical locations of the gates with respect to each other, and is related to the routing complexity of the circuit, and the capacity of each cluster.

In order to evaluate the physical locations of the gates, the *Cadence Virtuoso Placement and Route tool* is used to produce a compact floorplan layout from the schematic entry. Once the compact layout is constructed, the X,Y coordinates for every gate are extracted, and the cost functions are evaluated. Figure 4.19 presents the floorplan layout for the 4-bit CLA adder. The V_{dd} and *ground* rails are shown, and a cavity exists where the sleep transistors are located. The cavity has been taken into consideration when the X,Y coordinates of every gate are extracted.

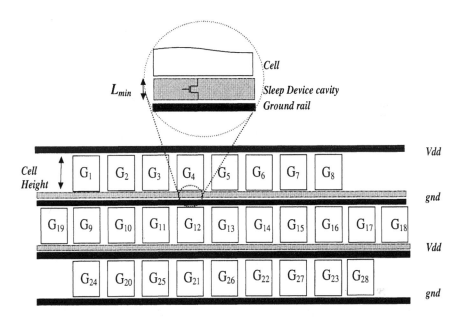

Figure 4.19. 4-bit CLA adder floorplan

In Figure 4.19, gates G_1 to G_{28} are identified, and the relative distances are computed from the compact layout. The cost function is formulated as follows:

$$c_j = (w_1 \times c_{j1}) + (w_2 \times c_{j2}) \tag{4.26}$$

where c_{j1} is a distance function (i.e., the rectilinear distance between the gates within a cluster) and c_{j2} represents the difference between the maximum cluster capacity and the sum of all the currents of the gates within a cluster. Therefore,

$$c_{j1} = \sum d_{uv} \text{ in a group } S_j \tag{4.27}$$

where d_{uv} is the distance between the centers of gates G_u and G_v. For example, in Figure 4.20, group S_j is composed of gates G_u, G_v and G_w. The value of the partial cost function of group S_j is $c_{j1} = d_{uv} + d_{vw} + d_{wu}$.
and

$$c_{j2} = Sleep\_Transistor_{max\_current} - \sum current_i \quad \forall i \qquad (4.28)$$

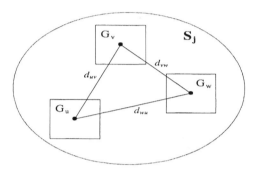

Figure 4.20.   Cost function calculation example

w_1 and w_2 are the weights associated with the cost of the two constraints, that is, the distance and capacity of the formed clusters. In this book, equal values are assigned to the weights $w_1=w_2=0.5$ in order to balance the distance and capacity constraints. The gates are grouped, and meet the constraint that the sum of currents does not exceed $I_{max}=250\mu A$. Figure 4.21 presents a very fast and efficient heuristic to form groups of clusters that will be used by the SP technique. The heuristic forms different types of clusters (i.e., clusters consisting of single gates, two gates, ...). This guarantees that the SP technique will find a solution for the problem. The target is to select certain clusters to achieve lowest cost value, while the I_{max} constraint is maintained. The groups must also cover all the gates with no repetition. As illustrated in Figure 4.21, the algorithm calculates the distance between each gate, and creates clusters consisting of single gates (this guarantees a solution to the BIP). In step (4) of the algorithm, the subroutine ($Create\_n\_Gate\_Clusters(cl)$) is utilized for creating clusters of different sizes (according to the parameter **cl** passed). In effect, a certain number (i.e., target) of clusters with a specific capacity is created, according to a parameter set within the algorithm. Currently, a limit of ten different clusters of a certain type is set as the upper bound. This number is set empirically to have a balance between the solution quality and the CPU time spent to solve the BIP problem.

The mathematical formulation of the Set Partitioning problem is as follows:

$$Minimize \quad Z = \sum_{j=1}^{n} c_j S_j \qquad (4.29)$$

subject to

$$\sum_{j=1}^{n} a_{ij} S_j = 1 \quad i = 1, ..., m \qquad (4.30)$$

$$S_j \in 0, 1 \quad j = 1, ..., n \qquad (4.31)$$

and

$$S_j = \begin{cases} 1, & \text{if the } j\text{th cluster is selected} \\ 0, & \text{otherwise} \end{cases}$$

where n is the number of groups generated, and $a_{ij} = 0$ or 1. In this formulation, each row ($i = 1, ..., m$) represents a constraint where module m should belong. The columns ($j = 1, ..., n$) represent feasible clusters (i.e., sleep transistors) that accommodate a set of gates in the circuit. The matrix a_{ij} is constructed as

$$a_{ij} = \begin{cases} 1, & \text{if gate i is covered by cluster j} \\ 0, & \text{otherwise} \end{cases}$$

Therefore, the objective of the low-power Set Partitioning problem is to find the *best* collection of clusters such that each gate is covered by exactly one cluster. The SP model is also a "0-1" pure integer LP problem which is again solved using CPLEX version 7.5.

Figure 4.22 portrays the solution of the SP technique with $w_1=w_2=0.5$ by highlighting the gates that are clustered together with the same pattern or color. It is also evident from the figure that the gates that are placed closely together are clustered (i.e., the gates in two consecutive rows) with a specific sleep transistor, thereby minimizing the wire-length.

To further illustrate how the floorplan of the clustered gates changes with different w_1, w_2 values, Figure 4.23 depicts the floorplan for $w_1=0.9$ and $w_2=0.1$. Since the distance dependent weight w_1 is given a larger value compared to the capacity dependent weight w_2, it can be seen from Figure 4.23 that the gates within a cluster are adjacent to each other, whereas the number of clusters has increased. This indicates that the efficiency of packing gates in a cluster has greatly degraded. In this case, the Set Partitioning modeling of the problem favors the *minimum distance* clustering over the *full capacity clustering*.

```
              CLUSTERING HEURISTIC
Create_Clusters()
1. Calculate distances between all gates;
2. Initialize maxgates_per_cluster=n;
3. Create clusters with Single gates;
4. For cl=2; cl ≤ maxgates_per_cluster
        Create_n_Gate_Cluster(cl)
     End For
5. For all clusters created calculate_cost()
6. Return();


Create_n_Gate_Clusters(cl)
1. For cluster of type cl
        create_new_cluster()
     While not done
          Choose Gate with minimum distance
          If sum of currents ≤ capacity
             append gate to newly created cluster
          End If
          If total gates within cluster ≥ limit
             break;
     End While
   End For
```

Figure 4.21.   Heuristic for Grouping Gates into Clusters

Furthermore, Figure 4.24 shows the normalized $P_{dynamic}$ and $P_{leakage}$ dissipation as a function of w_1, w_2, and the two curves are normalized to their minimum value. Minimum $P_{dynamic}$ is achieved at $w=(w_1=0.6, w_2=0.4)$, whereas minimum $P_{leakage}$ is achieved at $w=(w_1=0.5, w_2=0.5)$.

In general, the minimum power dissipation takes place at a balanced *minimum distance - full capacity* clustering. On the other hand, if the distance constraint is favored over the capacity constraint (i.e., $w_1 \gg w_2$), a large $P_{dynamic}$ and $P_{leakage}$ dissipation occurs. Therefore, equal w_1 and w_2 values are taken throughout this work to balance the *minimum distance* to *full capacity* clustering constraint.

Results and Discussion

Table 4.5 compares the SP and BP techniques in the literature. All the results are normalized to [7]. The BP and SP techniques employ sleep transistors which are sized to achieve a minimum FOM (as shown in Figure 4.18). A 5%

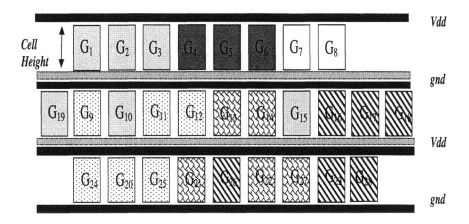

Figure 4.22. Results:4-bit CLA Adder Floorplan w=(0.5,0.5)

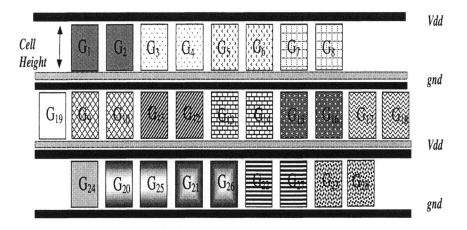

Figure 4.23. Results:4-bit CLA Adder Floorplan w=(0.9,0.1)

degradation in circuit speed is achieved, where the frequency of operation is set at 500MHz. The LVT and HVT are set to 350mV and 500mV, respectively in the 0.18μm CMOS technology. Furthermore, the interconnects to link the sleep transistors with the gates are taken into consideration. The area covered by these interconnects as well as the sleep transistor area, are reported. The $P_{dynamic}$ (P_d) and $P_{leakage}$ (P_l) are measured, while taking these intercon-

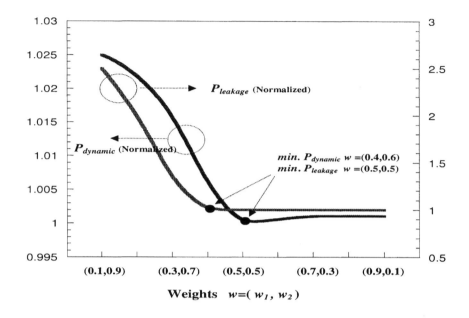

Figure 4.24.   Impact of weights over $P_{dynamic}$ and $P_{leakage}$

nects into account. Also, the capacitances associated with the metal interconnect lines and the parasitic cross-coupling capacitance are measured. When
the Metal 1 and Metal 2 lines are used, the capacitance for Metal 1 is 0.2236
fF/μm, whereas the capacitance for Metal 2 is 0.1905 fF/μm.

In Table 4.5, with the 5% speed degradation as the comparison basis, it is
clear that [4] employs a smaller sized single sleep transistor containing the
whole circuit compared to [7]. Consequently, a slight reduction in the dynamic
power is observed due to the reduction of the drain capacitance linked to the
sleep transistor. The results in [4] achieve an average of a 50% reduction in
leakage power compared to [7].

The highest leakage reduction occurs in the 27-bit CIC benchmark, due to
the large reduction in sleep transistor area from 3247 to 153. It should be em-
phasized that the CIC benchmark employs many gates that have mutually exclusive discharge patterns which enhance the efficiency of [4] (unlike the other
five benchmarks). However, the BP technique produces large reductions in the
sleep transistors' total area. Although the number of sleep transistors is higher

Table 4.5.   Algorithm Comparison

REF	Benchmark	4-bit CLA Adder	32-bit Parity Checker	6-bit Multiplier	4-bit 74181 ALU	32-bit Single Error Correcting C499	27-channel interrupt controller circuit C432
	No. of gates	28	31	30	61	202	160
[7]	Delay (Norm.)	1	1	1	1	1	1
	P_d (Norm.)	1	1	1	1	1	1
	P_l (Norm.)	1	1	1	1	1	1
	# Sleep Transistors	1	1	1	1	1	1
	Total ST_Area [$W_{sleep}(\mu m)$]	50	42	65	97	176	3247
[4]	Delay (Norm.)	1	1	1	1	1	1
	P_d (Norm.)	0.98	0.97	0.89	0.97	0.99	0.98
	P_l (Norm.)	0.58	0.51	0.23	0.41	0.46	0.05
	# Sleep Transistors	11→1	16→1	5→1	37→1	32→1	52→1
	Total ST_Area [$W_{sleep}(\mu m)$]	29.3	21.6	15	39.5	81	153
BP	Delay (Norm.)	1	1	1	1	1	1
	P_d (Norm.)	0.86	0.81	0.67	0.81	0.80	0.98
	P_l (Norm.)	0.066	0.062	0.041	0.072	0.052	0.0062
	P_d savings to [7]	14%	18.4%	31.4%	17%	20%	2%
	P_d savings to [4]	12.2%	15.9%	23%	14.4%	19.2%	0%
	P_l savings to [7]	93.4%	92.3%	89.0%	92.8%	94.8%	99.4%
	P_l savings to [4]	88.6%	84.8%	71.0%	82.4%	88.7%	87.6%
	# Sleep Transistors	3	4	3	7	5	11
	$W_{sleep}(\mu m)$	1.1	0.66	1.32	0.88	1.54	1.54
	Total ST_Area [$W_{sleep}(\mu m)$]	3.6	2.82	4.36	7.1	9.3	20.5
SP	Delay (Norm.)	1	1	1	1	1	1
	P_d (Norm.)	0.91	0.88	0.81	0.85	0.91	0.98
	P_l (Norm.)	0.117	0.12	0.15	0.126	0.13	0.0107
	P_d savings to [7]	7%	9%	19%	11%	9%	2%
	P_d savings to [4]	5.1%	6.2%	9%	8.2%	8.1%	0%
	P_l savings to [7]	87%	85%	85%	86%	87%	98.9%
	P_l savings to [4]	77.7%	70.4%	34.8%	65.6%	71.1%	76.6%
	# Sleep Transistors	9	9	9	18	22	40
	$W_{sleep}(\mu m)$	0.66	0.66	1.1	0.66	1.1	0.88
	Total ST_Area [$W_{sleep}(\mu m)$]	7.2	7.3	11.2	13.15	26.55	38.8

than [7] and [4], the size of each sleep transistor is much smaller, achieving an overall reduction in sleep transistor area. Therefore, the BP technique offers significant dynamic power savings compared to [7] and [4] as reported in Table 4.5. On average, the BP technique achieves a 17% reduction over that of [7], and a 14% dynamic power reduction over that of [4]. The main savings, though, is associated with the leakage power due to the reduction of the sleep transistor size, which is directly proportional to the leakage power dissipation. On average, the BP technique achieves a 85% and 95% leakage power reduction compared to that of [4] and [7], respectively.

The SP technique is then compared to the BP technique, the work in [7] and the work in [4], while still keeping the 5% speed degradation as the comparison basis. The SP technique produces large reductions in the sleep transistors' total area compared to that of [7] and [4], but is higher than that of BP. This occurs because an additional constraint to the objective function is added (i.e., routing cost), and no preprocessing is incorporated as explained earlier in this section. The SP technique reduces the dynamic power, on average, by 10% and 6% compared to [7] and [4], respectively. This is attributed to the reduction of the capacitance due to the down-sizing of the sleep transistors.

Furthermore, the SP technique achieves a 88% and 66% leakage reduction compared to that of [7] and [4]. The main advantage of the SP technique is that it takes into consideration the location of the blocks in order to reduce the overall interconnects, providing more optimization to the area. The advantages of the SP technique will be even more evident in the DSM regime, when interconnects dominate the circuit performance [8] and dynamic power. Moreover, equally sized sleep devices for a given benchmark, such as for BP or SP, facilitate design for other circuits and provides more regular layouts. The area of the sleep transistor (ST) is equal to $W_{sleep} \times L_{sleep}$. If the length of the sleep transistor (L_{sleep}) is kept constant in the four techniques mentioned in Table 4.5, the sleep transistor width (W_{sleep}) can now be used to represent the sleep transistor area. However, the total ST_Area values shown in Table 4.5 include the area of the sleep transistor interconnects, in addition to the sleep transistor itself.

The reduction in dynamic power is dependent on the number and size of the sleep transistors and how big the circuit is (the ratio of the sleep transistor capacitance to the overall circuit capacitance), whereas the leakage power is dependent on only the number and size of the sleep transistors. Therefore, it is evident in Table 4.5 that the savings in leakage power is proportional to the reduction in the total sleep transistor area. Finally, the proposed technique offers a minimal area overhead with no perturbation to the layout. This is attributed to the very narrow cavity (Figure 4.19)which is located at a fixed location parallel to either the supply or *ground* rails, that holds the sleep transistors. This further guarantees that the sleep transistor will not change the overall floorplan of the circuit. Another point that should be mentioned is that the discharge current at the output of a gate differs very little after the insertion of the *on* sleep transistor. An interesting highlight is that after applying the BP or SP techniques, some sleep transistors may still have the capacity to contain more gates (i.e., not fully utilized). Thus, these sleep transistors can be sized down, which would further reduce the leakage power. This can be investigated in future work to produce even more accurate results. However, the optimization steps will not change.

In summary, the proposed gate clustering technique in this work is fundamentally better than that in [7] and [4] because: (1) partially overlapping currents are taken into account, (2) more advanced heuristics are used, (3) the technique gives good results for the general structure of circuits, not only tree-shaped architectures, and (4) routing complexity is taken into account, which is an important issue in the DSM regime.

In order to further improve the results achieved, hybrid heuristics that combine the characteristics associated with the BP and SP algorithms, are developed.

4.9. Hybrid Heuristic Techniques

In this section, several hybrid heuristics to improve the performance of the Set Partitioning technique explained in the previous section are introduced. One of the bottlenecks in the Set Partitioning technique is the limitation of the clustering heuristic (introduced in Figure 4.21) to produce all the possible types of clusters that can be used by the BIP solver, that is, the clustering heuristic appends gates that are closest to form a cluster. There is no consideration of the overlapping current (i.e., the preprocessing of the gate currents) which was introduced in Section 4.8. The hybrid heuristics make use of the knowledge gained from the preprocessing of the gate currents, in addition to the closeness of gates, to form an effective cluster.

Figure 4.25 introduces the first heuristic technique H_A. For each I_{EQ} cluster formed (i.e., the set of gates with overlapping currents as in Section 4.8), a new cluster is created. All the gates that are close to the I_{EQ} cluster will be appended to the current cluster, as long as the capacity of the sleep transistor

---

HYBRID H_A

1. For (each I_{EQ} Cluster formed)
   Create a New Cluster (add all existing gates)
   Set Total Current to Max Current of I_{EQ} Cluster
   For (each gate closest to current formed cluster)
      If (gate does not belong to current cluster)
         TotalCurrent = I_{EQ}.MaxCurrent + Gate.Current
         If (TotalCurrent $\leq$ BinCapacity)
            Append Gate to Newly formed cluster
         else
            Reject the gate
      End If
      Update Cost for Newly created Cluster
   End For
   End For
2. Create SP Formulation of the benchmark
3. Solve the SP using CPLEX

Figure 4.25.   A Simple Preprocessing/SP Hybrid Heuristic

is not exceeded. For all the clusters formed, a new Set Partitioning is formulated and then solved by CPLEX as a BIP problem. The main advantage of this heuristic is that the formed clusters can have much more current capacity than the simple clustering heuristic introduced in Section 4.8. Solving a Set Partitioning problem based on this hybrid heuristic achieves results that utilize fewer sleep transistors than the pure Set Partitioning formulation based on the simple clustering technique. To further improve the performance of the Set Partitioning problem, a second hybrid technique H_B is presented. This hybrid not only merges the existing adjacent gates to the I_{EQ} clusters formed by the preprocessing, but also merges the I_{EQ} clusters.

As seen in Figure 4.26, the heuristic merges every I_{EQ} cluster with all the other possible I_{EQ} clusters. It is noteworthy that the efficiency of the heuristic will be less effective than the technique in Figure 4.21. It is expected that the solution quality of the two hybrid techniques is between those obtained by the Bin Packing formulation, and those obtained by the Set Partitioning problem (using a simple clustering heuristic).

The third hybrid heuristic H_C (shown in Figure 4.27) is similar to H_B except that all the gates close to the newly formed cluster (i.e., the combination of the I_{EQ} clusters) are appended, as long as their current does not exceed the

---

HYBRID H_B

1. **For** (each I_{EQ} Cluster formed)

    Find all possible I_{EQ} Candidates that can be added

    For (all possible new clusters that need to be created)

        Copy I_{EQ} Cluster to New Cluster (add all existing gates)

        Set Total Current to Max Current of I_{EQ} Cluster

        For (every other I_{EQ} Cluster in the pool)

        TotalCurrent = I_{EQ}.MaxCurrent + OtherI_{EQ}.MaxCurrent

        If (TotalCurrent $\leq$ BinCapacity)

          Append All Gates of New I_{EQ} to Newly formed cluster

        else

          Reject the I_{EQ}

        End If

    Update Cost for Newly created Cluster

    End For

    End For

2. Create SP Formulation of the benchmark
3. Solve the SP using CPLEX

Figure 4.26. An Effective $I_{EQ}+I_{EQ}$ SP Clustering Hybrid Heuristic

upper limit of the sleep transistor.

Finally, there is the hybrid heuristic technique H_D which creates clusters by utilizing all the previously explained hybrid heuristic approaches, in addition to the regular clustering heuristic introduced in Section 4.8.

Table 4.6 compares the results obtained by the four hybrid heuristics. It is clear from the table that the quality of the solutions obtained by the hybrid heuristics is better than that of the simple clustering technique initially proposed for solving the Set Partitioning problem (Table 4.7). It is also evident from Table 4.6 that combining the I_{EQ} clusters reduces the total number of sleep transistors (especially the number of sleep transistors with large capacities) but at the expense of the routing complexity. It is interesting that the hybrid heuristic technique H_D achieves similar results to those obtained by using the hybrid H_B, in terms of the number of sleep transistors utilized. An important fact that may be overlooked in this case is that the heuristic H_D accounts for the routing complexity of the gates to the sleep transistor. In addition, since hybrid H_D creates clusters by utilizing hybrids H_A, H_B and H_C, the Set Partitioning problem is less constrained (i.e., more clusters are formed) and therefore, can be solved in less time.

HYBRID H_C

1. **For** (each I_{EQ} Cluster formed)
       Find all possible I_{EQ} Candidates that can be added
       For (all possible new clusters that need to be created)
           Copy I_{EQ} Cluster to New Cluster (add all existing gates)
           Set Total Current to Max Current of I_{EQ} Cluster
           For (every other I_{EQ} Cluster in the pool)
               TotalCurrent = I_{EQ}.MaxCurrent + OtherI_{EQ}.MaxCurrent
               If (TotalCurrent $\leq$ BinCapacity)
                   Append All Gates of New I_{EQ} to Newly formed cluster
               else
                   Reject the I_{EQ}
           End If
           End For
           For (all other close gates to newly formed cluster)
               TotalCurrent = CurrentCluster.MaxCurrent + Gate.Current
               If (TotalCurrent $\leq$ BinCapacity)
                   Append the Gate to Newly formed cluster
           End For
       Update Cost for Newly created Cluster
       End For
       End For
2. Create SP Formulation of the benchmark
3. Solve the SP using CPLEX

Figure 4.27.   An Effective $I_{EQ}+I_{EQ}$+GATE Preprocessing/SP Clustering Heuristic

---

Hybrid H_D

1. Initialize
2. Create 1 gate clusters
3. Use Preprocessing Heuristic
4. Use Regular Clustering Heuristic
5. Use Hybrid H_A to create clusters
6. Use Hybrid H_B to create clusters
7. Use Hybrid H_C to create clusters
8. Remove redundant clusters
9. Create SP Formulation of the benchmark
10. Solve the SP using CPLEX

Figure 4.28.   Combined Hybrid Heuristics

Table 4.7 compares the results obtained by: Bin Packing, Set Partitioning based on simple clustering, and Set Partitioning based on the introduced hybrid heuristic H_D. It is clear from the table that the bin packing model pro-

Table 4.6. Comparison of Sleep Transistors For Hybrid Heuristics

Circuit	$I_{sleep}=150$				$I_{sleep}=250$				$I_{sleep}=350$				$I_{sleep}=400$			
	H_A	H_B	H_C	H_D	H_A	H_B	H_C	H_D	H_A	H_B	H_C	H_D	H_A	H_B	H_C	H_D
CLAD	9	8	9	7	6	5	5	5	5	4	5	4	5	3	5	3
Mult	10	10	10	10	9	6	6	6	6	3	3	3	6	2	3	2
Parity	8	4	4	4	5	3	3	3	4	2	4	2	4	2	4	2
Alu	17	17	17	17	10	6	7	6	9	5	6	5	8	4	5	4
Error	34	14	14	14	28	9	13	9	28	6	8	6	18	6	8	6
AllCh	53	47	47	47	32	19	22	19	25	12	13	12	23	11	11	11

Table 4.7. BP/SP/Hybrid Comparison of Sleep Transistors

Circuit	$I_{sleep}=150$			$I_{sleep}=250$			$I_{sleep}=300$			$I_{sleep}=350$			$I_{sleep}=400$		
	BP	SP	H_D	BP	SP	H_D	BP	SP	H_D	BP	SP	H_D	BP	SP	H_D
CLAD	6	9	7	3	6	5	3	6	4	3	5	4	3	6	3
Mult	6	18	10	3	9	6	2	9	2	2	8	3	2	7	2
Parity	4	9	4	3	7	3	2	6	2	2	6	2	2	6	2
Alu	10	18	17	6	12	6	5	11	5	4	11	5	4	11	4
Error	12	33	14	8	22	9	6	20	6	5	21	6	5	21	6
AllCh	32	56	47	16	33	19	13	30	14	11	28	12	10	27	11

duces solutions with the least number of sleep transistors. The Set Partitioning model, based on the simple clustering technique (introduced in Figure 4.21), gives the maximum number of sleep transistors but has a less complex routing. Finally, the results obtained from the Set Partitioning model (by using the hybrid heuristic technique) offers solutions that are balanced between the total number of sleep transistors to be used and the routing complexity. Since H_D has been the heuristic technique of choice, H_D will be denoted by the *Hybrid Problem* (HP).

Figure 4.29 displays the CPU time involved in solving the benchmarks for BP, SP (based on the simple clustering technique), and SP (based on the Hybrid H_D), respectively. It is evident from the figure that solving the SP problem involves more CPU cycles than solving the BP problem. This is due to the fact

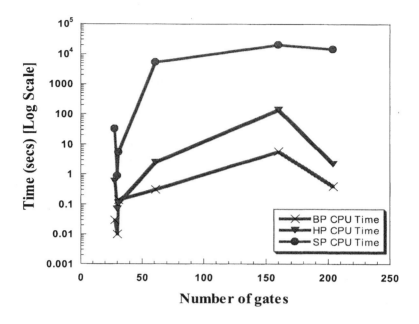

Figure 4.29. Computation time for BP, SP and HP

that the number of variables and constraints in the SP problem are much larger than that of the BP problem. The CPU time of the SP technique improves dramatically when the hybrid heuristic versus the simple clustering technique is invoked. This occurs because as the number of clusters generated for the SP technique is increased, the smaller the amount of computation time involved (as evident from the graph with respect to the hybrid approach). Figure 4.30 plots the effect of the size of the bin (I_{sleep}) with respect to the computation time for the ALU benchmark. The BP preprocessing algorithm has a worst case complexity of $O(n^2)$, where n is the number of gates in the circuit; the SP algorithm complexity is $O(nk)$, where n is the number of gates in the circuit, and k is the maximum gates to be appended in a cluster. For large circuits, it is recommended that heuristic search techniques such as Genetic Algorithms and Tabu Search be used instead of the CPLEX solver. H_D is chosen as the hybrid heuristic technique to be verified by the six benchmarks. This is done, because the H_D technique creates clusters using all the other hybrid heuristic approaches, and thus employs fewer sleep transistors in the design (Table 4.6), leading to dynamic and leakage power savings.

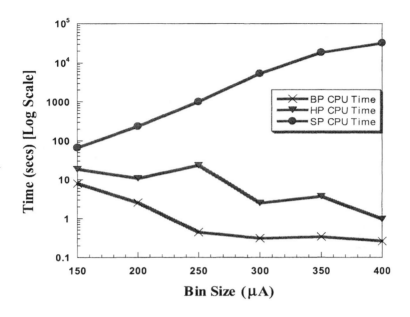

Figure 4.30. Computation Time for BP,SP and HP (ALU Benchmark)

4.10. Virtual Ground Bounce

So far, the criteria for sizing the sleep transistor have been performance (a 5% speed degradation is set), and the minimization of the dynamic and leakage power. However, an equally important design criterion is sizing the sleep transistor for noise. In MTCMOS circuits, virtual *ground* rails have a higher impedance than the true *ground* rails, and will, thus, unavoidably bounce. This will cause a serious reduction in gate speed as the effective supply voltage decreases, and a degradation in the noise margins. The problem with *ground* bounce is that many logic gates share a centralized sleep transistor, and hence, the same virtual *ground*.

Figure 4.31 shows the virtual ground bounce transient. It is generated by simulating the transient response of the virtual ground rail for different sleep transistor sizes for the ALU benchmark. The virtual *ground* rail attached to the sleep transistor that holds the highest number of gates is monitored. As the sleep transistor width decreases, the longer the time duration of the *ground* bounce bump. This is attributed to the RC time constant associated with the very high resistance case produced from the small sized sleep transistors. Con-

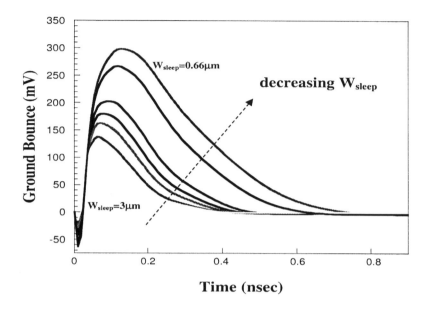

Figure 4.31.   Transient Response: Ground Bounce

sequently, this can cause the waveform at the gate output supported by that sleep transistor to slow down. Figure 4.32 portrays the variation of *ground* bounce according to the sleep transistor's size. The smaller the sleep transistor, the higher the *ground* bounce. Therefore, the sleep transistor should not only be sized for speed, and dynamic and leakage power, but for noise as well. Previous researchers [7], [4] have not included *ground* bounce in their analysis, a critical issue that should have been considered. Some physical issues related to the *ground* bounce will be illustrated, followed by the design methodology that takes ground bounce into account.

Impact of Virtual Ground Capacitance

The wire and junction capacitance associated with the virtual *ground* line should actually help reduce the *ground* bounce by serving as a local charge sink or reservoir for current [4] as shown in Figure 4.33. However, this capacitance would have to be extremely large in order to offset the effects of a poorly sized sleep transistor. The RC network serves as a low-pass filter, where the RC time constant must be large enough such that the virtual *ground* voltage can only rise to a fraction of its peak DC value.

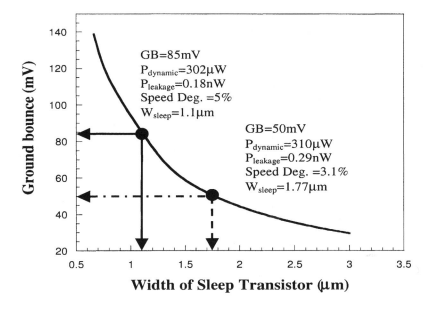

Figure 4.32. Ground bounce vs. W_{sleep}

If the time constant is very large, then it will also take longer for the virtual *ground* node to discharge back to the *ground* after a transition (see Figure 4.31). Rather than relying on large capacitances to ensure MTCMOS performance, it is much easier to lower the effective resistance with the appropriate sleep transistor size instead.

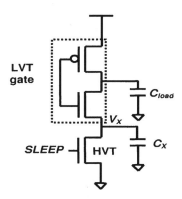

Figure 4.33. Capacitance associated with virtual *ground* rail

Reverse Conduction Paths through Virtual Ground

MTCMOS logic blocks can also have the disadvantage of reverse conduction as indicated in Figure 4.34, where current flows from the virtual *ground* though the LVT pulldown devices and charges up the output load capacitance [1]. The virtual ground rises above 0V so that another gate (which is supposed to be low) can experience reverse conduction, as the output voltage rises from 0 to V_X. This charging current comes from the discharging current of the other gates transitioning from high to low. As a result, the MTCMOS circuit is slightly faster, because the V_X voltage drop is not quite as large as expected if all the current flowed through the sleep transistor to the *ground*. Another effect of the reverse conduction is that the gate charging from low to high is faster, since it is already precharged to V_X. The drawback is that the noise margins in the circuits are reduced, and in the worst case, the circuit can fail logically. Therefore, the sleep transistor should again be properly sized to attain adequate noise margins.

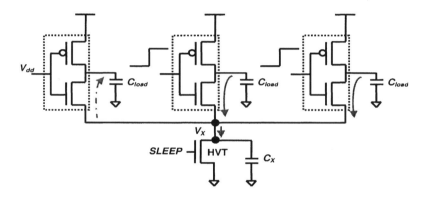

Figure 4.34. Reverse conduction

Design Methodology

In order to include *ground* bounce as a design criterion, dynamic and leakage power are reduced under two constraints that must be achieved simultaneously. First, the speed degradation is set to never exceed 5%, and secondly, the *ground* bounce is also set to never exceed 50mV. These constraints guarantee that the circuit will achieve sufficient speed and noise margins.

In Section 4.8, Table 4.4 reflects the values for W_{sleep} for a 5% speed degradation. For $W_{sleep} = 1.1\mu$m, the speed degradation due to the sleep transistor

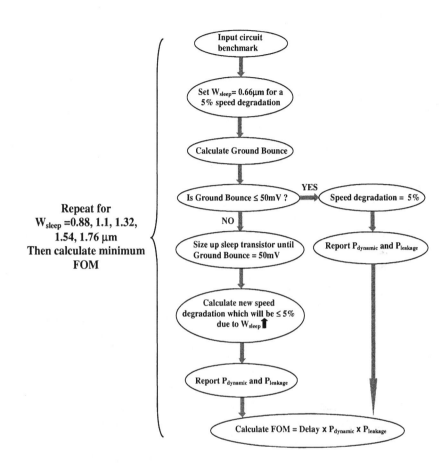

Figure 4.35. Design Methodology

is 5%, if I_{sleep}=250μA. Due to the small size of the sleep transistor both the dynamic and leakage power are reduced during the active and idle modes, respectively. However, the $ground$ bounce on the virtual $ground$ rail is high and equals 85mV which is over the acceptable noise limit (50mV). This is shown in Figure 4.32. As a result, the sleep transistor is sized up until the $ground$ bounce is reduced to 50mV. This is achieved at W_{sleep}=1.77μm. The sizing up of the sleep transistor actually enhances the circuit speed, and now causes only a 3.1% degradation in speed. However, the dynamic and leakage power can rise due to the sizing up of W_{sleep}. A figure of merit (FOM) is therefore established which takes into account delay, and dynamic and leakage power dissipation, while attaining a ground bounce $\leq$ 50mV.

$$Figure - of - Merit = P_{dynamic} \times P_{leakage} \times Speed_{Degradation} \quad (4.32)$$

This FOM is calculated for different sleep transistor sizes while adhering to the constraints, and the W_{sleep} achieving minimum FOM is recorded. It should be noted that the area associated with the sleep transistors is implicit in Equation (4.32) of the $P_{leakage}$ term.

The design methodology, represented by the flow diagram in Figure 4.35, is applied to each of the six benchmarks at six different sleep transistor sizes: W_{sleep} = 0.66, 0.88, 1.1, 1.32, 1.54, 1.76μm. The design methodology starts with the assumption that the speed degradation is equal to 5%. The *ground* bounce is then calculated. If *ground* bounce is less than or equal 50mV, then the speed degradation is taken as the pre-assumed 5%, and $P_{dynamic}$, $P_{leakage}$ are reported, followed by calculating the FOM; if the *ground* bounce is higher than 50mV, the sleep transistor is sized up until *ground* bounce = 50mV. The new value for the speed degradation is reported, as well as $P_{dynamic}$, $P_{leakage}$. The related FOM is then calculated. For every benchmark, the FOM is calculated for the different sleep transistor sizes, and the size achieving the minimum FOM is recorded. At this sleep transistor size, the speed, and $P_{dynamic}$ and $P_{leakage}$ powers are recorded. Since a single benchmark employs several sleep transistors, the *ground* bounce is always monitored on the virtual *ground* rail attached to the sleep transistor that holds the highest number of gates.

4.11.   Results: Taking ground bounce into account

Table 4.8 summarizes the results for the six benchmarks. The Bin Packing technique, taking ground bounce into account (BP_{GB}) is compared to the Bin Packing technique without taking ground bounce into account (BP) (Table 4.5). Similarly the Set Partitioning technique, taking ground bounce into account (SP_{GB}) is compared to the Set Partitioning technique without taking ground bounce into consideration (SP). Furthermore, the devised hybrid heuristic takes ground bounce into account HP_{GB}, and is compared to the BP, BP_{GB}, SP and SP_{GB} techniques.

The BP_{GB} technique achieves, on average 10% and 6% dynamic reduction compared to those of [7] and [4], respectively. Similar to BP, the main savings for BP_{GB} is associated with the leakage power, where 80% and 65% savings are achieved compared to those achieved in [7] and [4], respectively. From Table 4.8, the BP_{GB} technique achieves lower dynamic and leakage savings

Table 4.8.   Algorithm comparison taking ground bounce into account

REF	Benchmark	4-bit CLA Adder	32-bit Parity Checker	6-bit Multiplier	4-bit 74181 ALU	32-bit Single Error Correcting C499	27-channel interrupt controller circuit C432
	No. of gates	28	31	30	61	202	160
	Delay (Norm.)	1	1	1	1	1	1
	P_d savings to [7]	14%	18.4%	31.4%	17%	20%	2%
	P_d savings to [4]	12.2%	15.9%	23%	14.4%	19.2%	0%
BP	P_l savings to [7]	93.4%	92.3%	94.9%	92.8%	94.8%	99.4%
	P_l savings to [4]	88.6%	84.8%	77.8%	82.4%	88.7%	87.6%
	# Sleep Trans	3	4	3	8	6	13
	W_{sleep} (μm)	1.1	0.66	0.88	0.88	1.54	1.54
	Total ST_Area [$W_{sleep}(\mu$m)]	3.6	2.82	2.82	7.1	9.3	20.5
	Delay (Norm.)	1	1	1	1	1	1
	P_d savings to [7]	7%	9%	19%	11%	9%	2%
	P_d savings to [4]	5.1%	6.2%	9%	8.2%	8.1%	0%
SP	P_l savings to [7]	87%	85%	85%	86%	87%	98.9%
	P_l savings to [4]	77.7%	70.4%	34.8%	65.6%	71.1%	76.6%
	# Sleep Trans	9	9	9	18	22	40
	W_{sleep} (μm)	0.66	0.66	1.1	0.66	1.1	0.88
	Total ST_Area [$W_{sleep}(\mu$m)]	7.2	7.3	11.2	13.15	26.55	38.8
	Delay (Norm.)	0.61	0.644	0.13	0.24	0.163	1
	P_d savings to [7]	11%	16.5%	14.3%	9.4%	6.4%	2%
BP_{GB}	P_d savings to [4]	9.2%	13.9%	3.7%	6.6%	5.5%	0%
	P_l savings to [7]	90.5%	90.4%	69.2%	67.5%	99.3%	58%
	P_l savings to [4]	84.3%	81.2%	51%	25%	35.5%	98.2%
	# Sleep Trans	3	4	3	7	7	14
	W_{sleep} (μm)	1.77	1.005	6.675	4.485	7.97	1.32
	Total ST_Area [$W_{sleep}(\mu$m)]	5.84	4.42	22.03	34.5	61.4	20.3
	Delay (Norm.)	0.97	1	0.552	0.568	0.526	0.88
	P_d savings to [7]	9%	12%	12%	13%	8.7%	2%
SP_{GB}	P_d savings to [4]	7.1%	9.3%	1.1%	10.3%	7.8%	2%
	P_l savings to [7]	87.9%	85.9%	62%	78.8%	76.9%	98.8%
	P_l savings to [4]	79.1%	72.4%	38%	48.3%	49.8%	76%
	# Sleep Trans	9	9	9	18	33	40
	W_{sleep} (μm)	0.6713	0.66	2.74	1.142	1.23	0.984
	Total ST_Area [$W_{sleep}(\mu$m)]	6.65	6.55	26.8	22.86	44.95	43.25
	Delay (Norm.)	0.46	0.55	0.13	0.312	0.126	0.688
	P_d savings to [7]	10%	15%	14%	11.6%	7.7%	2%
HP_{GB}	P_d savings to [4]	8.2%	12.4%	3.4%	8.9%	6.8%	0%
	P_l savings to [7]	88.7%	88.8%	69.4%	80%	65%	99.1%
	P_l savings to [4]	80.5%	78%	45%	51.2%	24%	81.6%
	# Sleep Trans	5	4	3	8	13	20
	W_{sleep} (μm)	1.411	1.18	6.64	2.773	5.15	1.57
	Total ST_Area [$W_{sleep}(\mu$m)]	7.75	5.15	22.1	24.4	73.6	34.5

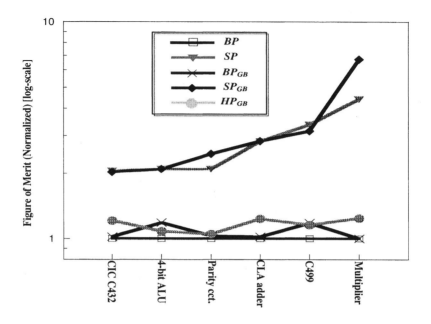

Figure 4.36. Comparison: Figure of Merit for benchmarks

compared to the BP approach. This is expected, since taking noise immunity into consideration causes the sleep transistor to be sized up, dissipating more dynamic and leakage power. However, due to this sizing up of the sleep transistor, a reduction in delay is associated with BP_{GB} compared to BP. By constructing the FOM (Equation(4.32)), it is evident from Figure 4.36 that BP_{GB} has slightly higher FOM values than those of the BP technique.

The SP_{GB} technique achieves, on average 9.5% and 6.3% dynamic reduction compared to [7] and [4] respectively. Furthermore, leakage power savings of 82% and 61% are achieved compared to that of [7] and [4], respectively. Similar to BP_{GB}, the SP_{GB} technique achieves lower dynamic and leakage savings than the SP approach does. This is again expected due to the same cause for the BP_{GB} case. Moreover, Figure 4.36 shows that both the SP and SP_{GB} achieve a high FOM. This is attributed to the large number of sleep transistors employed in the circuit due to the set-partitioning technique, which leads to large power values.

Although the HP_{GB} hybrid heuristic takes ground bounce into account, it achieves low FOM values. HP_{GB} achieves, on average, 10% and 6.6% dynamic reduction compared to that of [7] and [4], while 82% and 60% savings in leakage power are achieved compared to that of [7] and [4]. Figure 4.36 indicates that HP_{GB} achieves comparable FOM values to those of BP and BP_{GB}, but has the advantage that interconnect complexity is taken into consideration. Furthermore, HP_{GB} takes the noise associated with $ground$ bounce into consideration, unlike BP. Consequently, HP_{GB} can be considered the best technique for taking sleep transistor capacity constraints and interconnect complexity into account, while achieving low leakage, low dynamic power, and sufficient performance.

4.12. Power Management of Sleep Transistors

As mentioned in Chapter 2, most digital systems implemented in advanced CMOS technology are affected by the dissipation of leakage power in the active and standby modes; in particular, battery-operated wireless devices, where the standby time is much longer than the active processing time. For example, a cellular phone intermittently checks for incoming calls. During this intermittent operation, the standby time t_{off} is much longer than the processing time t_{on}. This is illustrated in Figure 4.37. To minimize the standby leakage power in the waiting time t_{wait}, the high-V_{th} sleep transistor must be controlled, to effectively control the MTCMOS sleep mode. A power management technique is popular, which supplies power to the digital system (DSP in the cellular phone) only when high-throughput processing is required.

The Power Management Processor

A power managment processor (PMP) effectively controls the sleep mode in MTCMOS circuits. Depending on the application and running sequence of the system (ratio of active to standby times), the PMP is programmed to control the power management sequence. The digital block (DSP) is implemented in a low-V_{th} structure, and is connected to the ground rail via the high-V_{th} sleep transistor, which is controlled by the PMP. On the other hand, the PMP is implemented with only high-V_{th} devices to eliminate standby leakage power that occur when the DSP is processing data. The PMP sends the $SLEEP$ signal that switches the high-V_{th} sleep transistor on and off. The PMP is shown in Figure 4.38.

Power Management Sequence

Figure 4.39 shows the power consumption by a DSP with and without power management. Without power management, standby leakage power is dissipated, whereas with power management, standby power is eliminated. This, a significant amount of battery energy is saved. During the power-on period

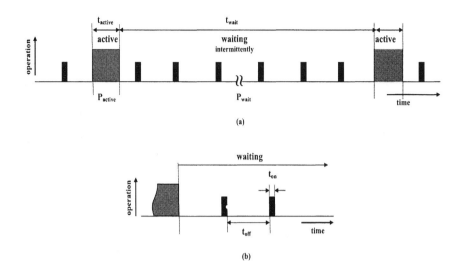

Figure 4.37. Conventional mobile phone running sequence: (a) the complete sequence and (b) detail of the waiting state. [15]

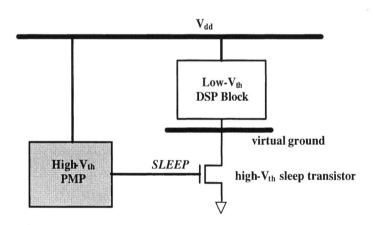

Figure 4.38. Power management processor

needed for checking, the PMP sets the $SLEEP$ signal $high$, allowing the DSP to process signals. After the check, the PMP sets the $SLEEP$ signal low again, turning off the DSP power supply. This limits the standby leakage current to that of the high-V_{th} sleep transistor. As a result, the standby power consumption is much lower than the conventional sequence with no power management. In addition, because the PMP is composed of high-V_{th} transistors and operates at a lower frequency than the DSP, the PMP power

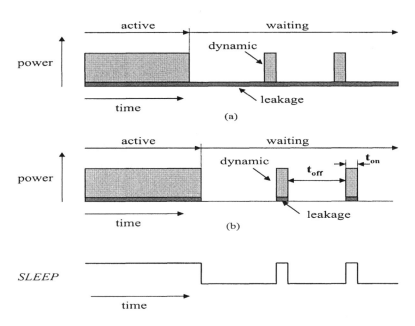

Figure 4.39. Power management sequences: (a) without power management, and (b) with power management [15]

4.13. Chapter Summary

The chapter gives an overview about MTCMOS designs employing sleep transistors. The sizing of these sleep transistors is done by three techniques: (1) the variable breakpoint switch level simulator, (2) hierarchical sizing based on mutually exclusive discharge patterns, and (3) the average current method. The drawbacks of these three techniques are illustrated, and solved by the distributed sleep transistor scheme devised by the authors. By applying the distributed sleep transistor scheme to six benchmarks to verify functionality, the results indicate that the proposed technique can achieve, on average 84% and 12% savings for leakage power and dynamic power, respectively, to existing techniques. In addition, the noise associated with $ground$ bounce is taken as a design parameter in the optimization problem.

Notes

1 Due to the approximation of the velocity saturation index α from 1.3 to 1 in
   the analysis, there is a small difference between the simulated current and
   the modeled current from Equation(4.15) which is in the order of 10%. This
   current difference may slightly shift the results later. However, the proposed
   clustering technique is not effected, and substantial leakage power savings
   are still achieved.

References

[1] J. Kao, A. Chandrakasan, and D. Antoniadis, "Transistor Sizing Issues And Tools For Multi-threshold CMOS Technology," in *Proceedings of the 34th Design Automation Conference*, 1997, pp. 409–414.

[2] Y. Ye, S. Borkar, and V. De, "A New Technique for Standby Leakage Reduction in High-Performance Circuits," in *Proceedings of the 1998 Symposium on VLSI Circuits*, June 1998, pp. 40–41.

[3] Z. Chen, L. Wei, and K. Roy, "Estimation of Standby Leakage Power in CMOS Circuits Considering Accurate Modeling of Transistor Stacks," in *Proceedings of the International Symposium on Low-Power Electronics and Design*, August 1998, pp. 239–244.

[4] J. Kao, S. Narendra, and A. Chandrakasan, "MTCMOS Hierarchical Sizing Based on Mutual Exclusive Discharge Patterns," in *Proceedings of the 35th Design Automation Conference*, 1998, pp. 495–500.

[5] J. Kao, *Subthreshold Leakage Control Techniques for Low Power Digital Circuits*, Ph.D. Thesis, Massachusetts Institute of Technology, May 2001.

[6] S. Mutah, S. Shigematsu, W. Gotoh, and S. Konaka, "Design Method of MTCMOS Power Switch for Low-Voltage High-Speed LSIs," *in Proceedings of Asia and South Pacific Design Automation Conference*, pp. 113–116, January 1999.

[7] S. Mutah, T. Douseki, Y. Matsuya, T. Aoki, S. Shigematsu, and J. Yamada, "1-V Power Supply High-Speed Digital Circuit Technology with Multi-Threshold Voltage CMOS," *IEEE Journal of Solid-State Circuits*, vol. 30, no. 8, pp. 847–853, August 1995.

[8] M. Bohr and Y. El-mansy, "Technology for Advanced High-Performance Microprocessors," *IEEE Transactions on Electron Devices*, vol. 45, no. 3, pp. 620–625, March 1998.

[9] M. Anis, S. Areibi, M. Mahmoud, and M. Elmasry, "Dynamic and Leakage Power Reduction in MTCMOS Circuits Using an Automated Efficient Gate Clustering," in *Proceedings of the 39th Design Automation Conference*, 2002, pp. 480–485.

[10] M. Anis, S. Areibi, and M. Elmasry, "Design and Optimization of Multi-Threshold CMOS (MTCMOS) Circuits," *IEEE Transactions on Computer-Aided Design of Integrated Circuits and Systems*, vol. 22, no. 10, October 2003 (To appear).

[11] A. Bellaouar and M. Elmasry, *Low-Power Digital VLSI Design Circuits and Systems*, Kluwer Academics Publications, 1995.

[12] Y. Taur and T. Ning, *Fundamentals of Modern VLSI Devices*, Cambridge University Press, 1999.

[13] D. Sylvester and C. Hu, "Analytical Modeling and Characterization of Deep-Submicron Interconnect," *Proceedings of the IEEE*, vol. 89, no. 5, pp. 634–664, May 2001.

[14] R. Rardin, *Optimization in Operations Research*, Prentice Hall, Boston, 1998.

[15] S. Mutoh, S. Shigematsu, Y. Matsuya, H. Fukuda, T. Kaneko, and J. Yamada, "A 1-V Multithreshold-Voltage CMOS Digital Signal Processor for Mobile Phone Applications," *IEEE Journal of Solid-State Circuits*, vol. 31, no. 11, pp. 1795–1802, 1996.

[16] J. Hatler and F. Najm, "A Gate-Level Leakage Power Reduction Method for Ultra Low-Power CMOS Circuits," in *Proceedings of the IEEE Custom Integrated Circuits Conference*, 1997, pp. 475–478.

Further Reading

- S. Kawashima, T. Shiota, I. Fukushi, R. Sasagawa, W. Shibamato, A. Tsuchiya, and T. Ishihara, "A 1V, 10,4mW Low Power DSP Core for Mobile Wireless Use," *IEICE Trans. Electron.*, VOL. E83-C, NO.11, pp. 1739-1749, November 2000.

- M. Stan, "CMOS Circuits with Subvolt Supply Voltages," *IEEE Design & Test of Computers*, VOL.19, NO.2, pp. 34-43, March-April 2002.

- T. Douseki, J. Yamada, and H. Kyuragi, "Ultra Low-power CMOS/SOI LSI Design for Future Mobile Systems," *in Symposium on VLSI Circuits Digest of Technical Papers*, pp. 6-9, June 2002.

- A. Agrawal, H. Li, and K. Roy, "DRG-Cache: A Data Retention Gated-Ground Cache for Low Power," *in Proc. ACM/IEEE Design Automation Conference*, pp. 473-478, June 2002.

- F. Hamzaoglu and M. Stan, "Circuit-Level Techniques to Control Gate Leakage for sub-100nm CMOS," *in Proc. IEEE International Symposium on Low Power Electronics and Design*, pp. 60-63, August 2002.

- K. Usami, N. Kawabe, M. Koizumi, K. Seta, and T. Furusawa, "Selective Multi-Threshold Technique for High-Performance and Low-Standby Applications," *IEICE Trans. Fundamentals*, VOL. E85-A, NO.12, December 2002.

- J. Tschanz, S. Narendra, Y. Ye, B. Bloechel, S. Borkar, and V. De, "Dynamic Sleep Transistor and Body Bias for Active Leakage Power Control of Microprocessors," *in International Solid-State Circuits Conference Digest of Technical Papers*, pp. 102-103, February 2003.

- B. Calhoun, F. Honore, and A. Chandrakasan "Design Methodology for Fine-Grained Leakage Control in MTCMOS," *in Proc. IEEE International Symposium on Low Power Electronics and Design*, August 2003. (To appear)

- H. Won, K. Kim, K. Jeong, K. Park, K. Choi, and J. Kong, "An MTCMOS Design Methodology and Its Application to Mobile Computing," *in Proc. IEEE International Symposium on Low Power Electronics and Design*, August 2003. (To appear)

Chapter 5

MTCMOS SEQUENTIAL CIRCUITS

5.1. Introduction

In the previous chapter, the MTCMOS technique was used to reduce standby leakage power in combinational logic circuits, while attaining a sufficient performance. However, special attention must be paid to the MTCMOS design of sequential logic circuits such as latches and flip-flops (FFs) that have memory functions; this is necessary because stored data in the latch or FF circuits cannot be retained during the sleep standby mode because the virtual $ground$ rails would be floating to cut off the leakage current. This would disconnect the storage nodes from the true power supply rails and possibly corrupt the stored data.

5.2. MTCMOS Latch Circuit

In 1996, Mutoh et al. proposed the first and most straightforward MTCMOS latch shown in Figure 5.1 [1]. Data is preserved through two parallel high-V_{th} CMOS inverters that provide static circulation during the standby mode. The two high-V_{th} CMOS inverters I_2 and I_3 are connected directly to the true power supply rails V_{dd} and $ground$, unlike I_1. Therefore, during the standby mode, the clock (CLK) is kept low, and the state of the latch is held through I_2 and I_3. Inverter I_3 is designed to be smaller to suppress the increase in the gate time delay and area.

Since latches are speed critical elements that make up a large fraction of the overall circuitry in modern digital designs, fast, low-V_{th} devices must be utilized in the forward path of MTCMOS latches. Therefore, inverters I_1, I_4 and the CMOS transmission gate TG_1 are composed of low-V_{th} devices (low-

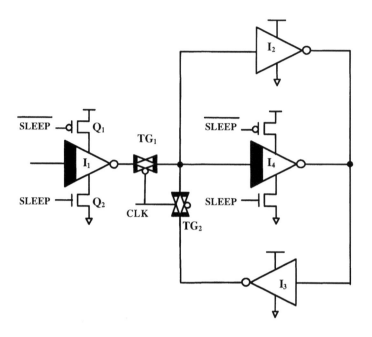

Figure 5.1.   Conventional MTCMOS latch

V_{th} structures are highlighted in Figure 5.1. The circuit is composed of local sleep transistors Q_1 and Q_2 with a high-V_{th} in order to cut off the leakage current paths. With the absence of the sleep transistors Q_1 and Q_2, a leakage current path would exist and flow through the PMOS of I_1 and I_3, caused by shorting V_{dd} and virtual V_{dd}.

5.3.   MTCMOS Balloon Circuit

Another MTCMOS circuit for preserving data during the sleep mode was proposed by Shigematsu et al. in 1997 [2]. It involves the use of a memory circuit which is always powered, in addition to a switch. The memory circuit is called a *balloon circuit*, because of its shape and because data is *blown* into the memory circuit at the start of the standby mode to preserve the data and to release it at the end to restore it.

Thus, this scheme has two states according to the circuit mode: standby and active. In the standby mode, the balloon circuit preserves data, and the leakage current path from the memory circuit to the logic circuit is cut off by the switch TG_2 (Figure 5.2). In the active mode, the balloon circuit does not add to the load of the logic circuit, because it is separated by the switch TG_2 (Figure 5.2). During the transitional time period between the active and standby modes, the switch TG_2 turns *on* to either read from or restore data to the logic circuit.

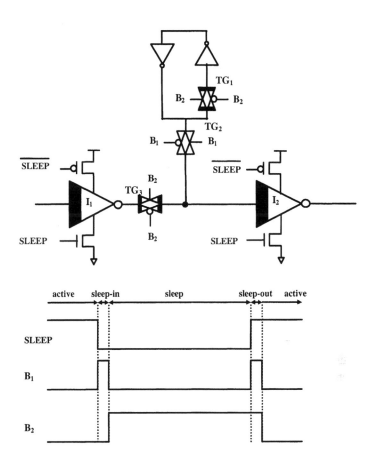

Figure 5.2. MTCMOS balloon circuit

The Balloon Circuit

The balloon circuit consists of two high-V_{th} CMOS inverters that are directly connected to the power supply rails and a transmission gate TG_1. The switch S is composed of a high-V_{th} transmission gate TG_2 to cut off the leakage current in the standby mode. Transmission gates TG_1 and TG_3 are assigned a low-V_{th}, because they are used to perform the delay-critical reading and writing to the memory circuit. These transmission gates are controlled by the signals B_1 and B_2 (Figure 5.3).

Figure 5.3 illustrates the operation of the balloon circuit. There are four periods in the logic circuits: active, standby, sleep-in, and sleep-out.

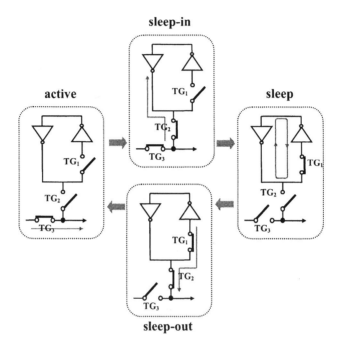

Figure 5.3. Operation of balloon circuit

- **Active period**: The balloon circuit is separated from the logic circuit by TG_2, which keeps the load down so as not to affect the speed. In this period, the balloon circuit does not operate.

- **Sleep-in period**: It is the transitional period from the active period to the sleep period. As suggested from its name, the data is read from the logic circuit and the circuit is ready to enter the sleep mode.

- **Sleep period**: The balloon circuit holds the data during that period. TG_2 cuts off the leakage current path from the balloon circuit.

- **Sleep-out period** It is the transitional period from the sleep period to the active period. Again, as suggested by its name, data is restored to the logic circuit and the circuit exits the sleep mode.

This operation enables the active-mode operation to resume smoothly, where the state of the balloon circuit changes only during the transitional periods between the sleep and active periods. Furthermore, the balloon circuit is isolated from the forward data path in the active period, such that the balloon circuit does not become a bottleneck during high-speed operation of the low-V_{th} application circuit. Therefore, the balloon circuit does not need to operate at a high speed, and can be designed with high-V_{th} and minimum size MOS devices. This allows lower standby power with a minimal increase in area.

A Balloon Master-Slave Flip-Flop

Generally, the Master-Slave Flip-Flop (MS-FF) is used for holding data. When the input clock CLK is $high$, the data is held in the master latch, and when the CLK goes low, the data is held in the slave latch. A typical CMOS MS-FF circuit is depicted in Figure 5.4. In the MTCMOS circuit, this MS-FF cannot preserve data during the sleep period because the virtual rails which are connected to the inverters, float.

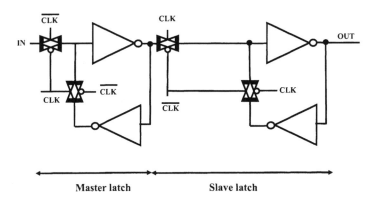

Figure 5.4. Typical CMOS MS-FF Circuit

This problem can be solved by utilizing the balloon circuit in the design of the MS-FF. The balloon circuit is used to preserve the data which the MS-FF holds. In this balloon MS-FF, the transmission gates TG_1, TG_2 and TG_3 are controlled by signals B_1 and B_2 in the same way as in the balloon circuit.

In the MS-FF circuit, the balloon circuit preserves only the data which is held in the master latch. When the low-V_{th} logic circuit block is designed, the balloon D-Flipflops (DFFs) are chosen according to the state of the CLK during the sleep-in and sleep-out periods. Because it is not necessary to preserve the state of the CLK, these balloon DFFs have only one balloon circuit. Therefore, they can suppress the increase in the circuit size, and lower the power to preserve data during the sleep period. An edge triggered MS-FF that utilizes the balloon storage mechanism is depicted in Figure 5.5.

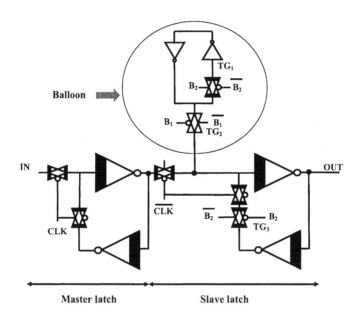

Figure 5.5.   Balloon Circuit Applied to DFF

Although the basic operation is simple, the balloon circuit approach requires a redundant circuitry that must be used for each MS-FF, and a complex timing methodology. This methodology achieves a clean abstraction between the balloon circuit and the MTCMOS logic block to eliminate any interaction between the circuits. In addition to the extra circuitry needed and the complex control, routing the control signals throughout the chip to each MS-FF can be costly in a large design. This problem would be more evident in the deep-submicron (DSM) regime, because of the timing constrained problems.

In order to eliminate the control and area costs of these sequential circuits, several alternative solutions have been proposed, such as the Intermittent Power Supply Scheme, the Auto-Backgate-Controlled MTCMOS, and the Virtual Clamp circuit.

5.4. Intermittent Power Supply Scheme

A low-power MTCMOS data storage circuit targeting sub-1V, was devised [3] [4] with a periodic refresh mechanism to save data, but there is extra overhead and potential noise issues.

The key feature of this scheme, the *Intermittent Power Supply (IPS) Scheme*, is that during the standby mode, the virtual ground rails VV_{dd} and VV_{ss} are connected before the data is damaged in the data storage circuit. This is done by refreshing the storage node by using a periodic refresh mechanism.

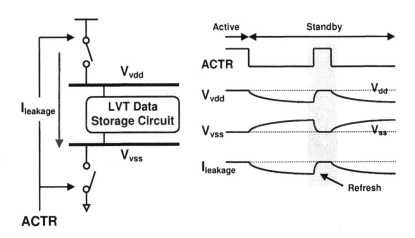

Figure 5.6. Concept of IPS

During the standby mode, the IPS scheme operates as follows: (1) the virtual power rails are cut off from the power supply lines, (2) the virtual power rails are pulled up (or pushed down) by the leakage current from the data latch circuit, (3) the off-leakage current is significantly reduced due to the establishment of a negative gate-to-source voltage, and (4) the virtual power rails are connected to the real power rails in a periodic fashion to refresh the stored data. Figure 5.6 shows the timing diagram of the IPS data storage circuit. The VV_{dd} and VV_{ss} levels depend on the ACTR signal. This technique, however, suffers from extra overhead and potential noise issues.

5.5. Auto-Backgate-Controlled MTCMOS

Two other methods of retaining the state in MTCMOS circuits were also devised in [5] [6]. In these approaches, the MTCMOS structure is modified with diode devices such that during the sleep mode, internal nodes do not float. As a result, the logic gates continue to operate on a reduced voltage swing, and the leakage currents are still reduced. However, the leakage reduction amounts are not as large as that in a conventional MTCMOS circuit, and there can be issues of robustness to insure functionality in a complicated circuit.

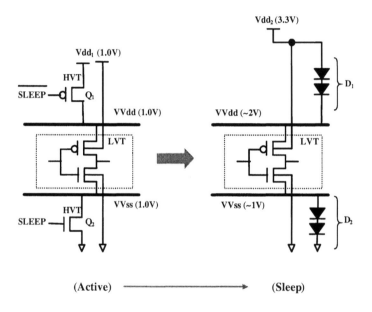

Figure 5.7.   Concept of an ABC MTCMOS gate

Figure 5.7 illustrates the concept of the Auto-Backgate-Controlled (ABC) MTCMOS circuit. In the active mode (SLEEP="1"), Q_1 and Q_2 are turned *on*, allowing the virtual voltage rails VV_{dd} and VV_{ss} to become Vdd_1 and 0V. Vdd_1 is taken to be 1.0V. Therefore, the internal low-V_{th} block operates with a low-power dissipation by the 1.0V power supply. In the sleep mode, Q_1 and Q_2 are turned *off*, and the voltage source switches to Vdd_2=3.3V. The potential of the virtual ground rails begins to rise due to the leakage current flowing from Vdd_2 and *ground*. This rise stops when the diode currents equalize the leakage current of the internal low-V_{th} block. The voltages of VV_{dd} and V_{ground} are approximately 2V and 1V, respectively. When the backgates of the device in the internal block become reverse-biased, the leakage current is greatly decreased. Because this circuit preserves all the levels of internal nodes, the latched data do not disappear.

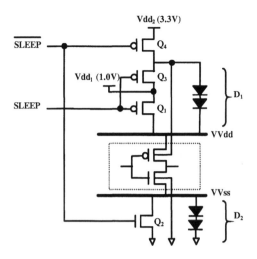

Figure 5.8. ABC MTCMOS gate

The ABC-MTCMOS scheme was applied to the logic core of a 32-bit RISC microprocessor and a 32Kbit gate array SRAM [6]. Then, an improved version of the ABC-MTCMOS scheme was presented in [7]. The improved designs allows for a reduction in the supply voltage, while retaining the data of the internal block, and reducing the leakage current significantly. Although sophisticated circuit techniques have been explored to maintain state during the standby mode, there is still a need for robust, fast, and efficient solutions.

5.6.   Virtual Rails Clamp (VRC) Circuit

The Virtual Rails Clamp (VRC) technique [8], shown in Figure 5.9, solves the data holding problem during the standby mode. Similar to MVCMOS, the sleeping transistors are low-V_{th} devices. These transistors are turned off in the idle periods so that the virtual lines will float. Due to the leakage current, the VV_{dd} line will begin dropping below V_{dd}, and the VV_{ss} line will start to rise above $ground$. Once the voltage on the VV_{dd} line drops by an amount equivalent to the diode's built-in potential: ϕ (ϕ=0.5V), the diode is turned on, thereby preventing any further discharge.

Similarly, any voltage increase on the VV_{ss} line will be compensated for by the other clamping diode. The difference between the two virtual lines, $V_{dd} - 2\phi$, makes it possible to hold the data during the sleep mode. In that mode, the leakage current is reduced, because the voltage drop increase on

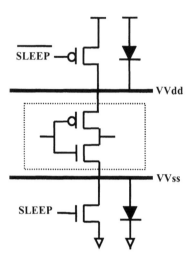

Figure 5.9.   Virtual Rails Clamp (VRC)

the virtual lines causes an increase in the bulk-to-source voltages of internal devices. This in turn, increases the threshold voltages of the sleep transistors (body effect), while negatively biasing the gate-to-source voltages of the internal transistors. However, unlike the MTCMOS scheme, the VRC approach results in leakage current between the true and virtual rails due to the use of low-V_{th} sleep transistors. Also, the VRC approach sets a lower limit on V_{dd} in order to have a sufficient logic swing ($V_{dd} - 2\phi$) to guarantee the preservation of the data. The VRC scheme saves leakage power in the standby mode due to the *stacking* of turned-*off* transistors. Having more than one turned-*off* transistor in a stack creates an intermediate voltage on the parasitic capacitance between the transistors, which causes an increase in the effective V_{th} of the transistors in the stack (the DIBL effect).

5.7.   Leakage Sneak Paths in MTCMOS Sequential Circuits

Kao et al. indicated that a problem with MTCMOS sequential circuits that utilize feedback and parallel devices is that sneak leakage paths may exist during the standby state [9]. This causes large leakage power dissipation in the standby mode.

In Chapter 4, it has been illustrated that a single high-V_{th} sleep transistor of either polarity is sufficient to cut off standby leakage currents in MTCMOS combinational logic blocks. No sneak leakage paths exist in these configurations, because all paths from V_{dd} to $ground$ must pass through the off high-V_{th} sleep transistor. However, when CMOS gates and MTCMOS gates are combined in sequential circuits, sneak leakage paths can arise by bypassing the off high-V_{th} transistors. Typically, such paths arise whenever the output of an MTCMOS gate is connected to the output of a CMOS gate.

There are two causes of sneak paths in MTCMOS sequential circuits: (1) single-polarity high-V_{th} devices, and (2) CMOS-MTCMOS reverse conduction paths.

Sneak leakage paths due to single-polarity high-V_{th} devices

Typically, there are two causes of sneak paths involving single-polarity high-V_{th} devices: (1) sneak leakage paths due to CMOS-MTCMOS parallel combinations, and (2) sneak leakage paths due to CMOS-MTCMOS connection through low-V_{th} passgates.

Sneak Leakage Paths due to CMOS-MTCMOS Parallel Combinations:

Figure 5.10 presents a configuration where sneak leakage paths can arise when a single polarity device is used. In this configuration, a CMOS gate is in parallel with an MTCMOS gate, and the output of the MTCMOS gate is connected to the output of the CMOS gate.

Consequently, for the single PMOS high-V_{th} device in Figure 5.10(a), the leakage sneak path flows through the on high-V_{th} PMOS of I_2 and the off low-V_{th} NMOS of I_1. Therefore, the sneak leakage path is restricted by the off low-V_{th} NMOS of I_1. For the single NMOS device case in Figure 5.10(b), the leakage sneak path flows through the on high-V_{th} NMOS of I_2 and the off low-V_{th} PMOS of I_1.

Therefore, dual-polarity high-V_{th} sleep devices should be used for the MTCMOS gates that are parallel with CMOS gates, where the output of the MTCMOS gate is also connected to the CMOS gate output in order to eliminate these leakage sneak paths.

Sneak leakage paths due to CMOS-MTCMOS connection through low-V_{th} passgates:

The second source for sneak leakage paths occurs when the output of an MTCMOS gate is connected to the output of a CMOS gate through low-V_{th} passgate devices. Whether the input of the latch is $high$ or low, large leakage

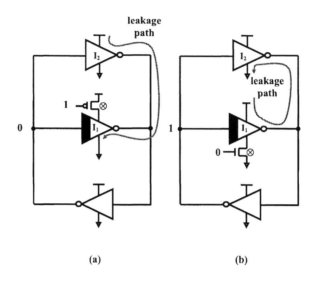

Figure 5.10.   Leakage paths due to single polarity sleep transistors

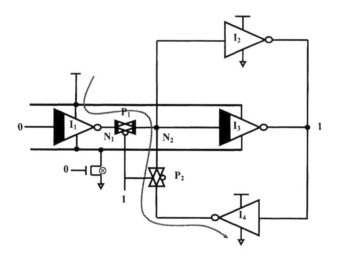

Figure 5.11.   Leakage paths through low-V_{th} passgate devices

currents will flow. This can be understood by examining the following two scenarios.

For a single polarity NMOS sleep transistor, if the latch input is driven *low*, and the latch output is held *high* in the standby state (Figure 5.11, the internal nodes N_1 and N_2 will be strongly driven to opposite polarities, and because passgate P_1 cannot turn *off* strongly, the result is large leakage currents which can flow through the turned *on* PMOS device of I_1, the turned-*off* low-V_{th} passgate P_1, the turned *on* high-V_{th} passgate P_2, and the turned *on* high-V_{th} NMOS of I_4.

On the other hand, if the input to the latch is *high* (same Figure 5.11), and the latch output is still held *high* there will be a sneak leakage path through the *off* low-V_{th} PMOS of I_1 in series with the *off* low-V_{th} passgate P_1, the *on* high-V_{th} passgate P_2, and the *on* high-V_{th} NMOS of I_4. In this case, the leakage path is the same, but because the leakage path will flow through two series *off* devices, there will be a reduction in leakage currents compared to the previous case. This leakage reduction is attributed to the de-biasing effect discussed in Chapter 2, but the reduction can still be an order of magnitude larger than necessary. Sneak paths which allow leakage sneak paths to exist, when a single polarity PMOS device is employed will now be examined (Figure 5.12).

When the latch input is driven *high*, and the output is holding a "0", then again the internal nodes N_1 and N_2 will be driven to opposite polarities, resulting in large leakage currents during the sleep mode. The sneaking leakage current is created from the turned *off* low-V_{th} P_1 passgate. The sneak leakage path will flow through the *on* high-V_{th} PMOS of I_4, the *on* high-V_{th} passgate P_2, the *off* low-V_{th} passgate P_1, and the *on* high-V_{th} NMOS of I_1.

It is evident that employing a second polarity high-V_{th} NMOS sleep transistor can cut off these leakage sneak paths. Therefore, both polarity sleep devices are required in inverter I_1 to eliminate this sneak leakage path completely. The drawback, however, is that sleep devices with both polarities require more area.

Sneak leakage paths due to CMOS-MTCMOS reverse conduction paths

It has been illustrated that sneak leakage current paths can be eliminated by using both polarity sleep devices. However, in some cases, simply using both polarity sleep devices does not guarantee the complete elimination of sneak leakage paths.

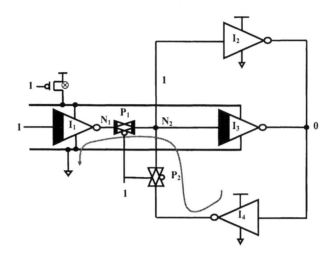

Figure 5.12. Leakage paths through single polarity Sleep transistors

As mentioned at the end of Chapter 4, reverse conduction leakage paths can exist if multiple MTCMOS gates share the same high-V_{th} sleep devices. Reverse conduction leakage paths can arise if the MTCMOS gates share the common virtual V_{dd} or virtual *ground* line, and if the MTCMOS-CMOS gate pairs have a common output node, or are connected through low-V_{th} passgates. An example of such a sneak path is presented in Figure 5.13.

The reverse conduction sneak path originates from V_{dd} supply rail of one of the CMOS gates, travels through the virtual power lines and exits to *ground* through another CMOS gate. Thus, the current flows through an MTCMOS gate in a reverse conduction path, that is, a PMOS current that flows from the device output up towards virtual V_{dd}, or a NMOS current that flows from virtual *ground* up to the device output.

For the reverse conduction sneak path, the only turned *off* device is the low-V_{th} passgate P_1 which causes the leakage current to be greater than the leakage current when the two low-V_{th} devices are turned *off*. All other devices in the leakage path are strongly turned *on*: the PMOS of inverter I_2, the PMOS of I_3, the PMOS of I_1, the passgate P_2, and the NMOS of I_4.

Kao et al. stated that in general, reverse conduction sneak leakage paths for a block with common virtual power or *ground* lines can be eliminated by en-

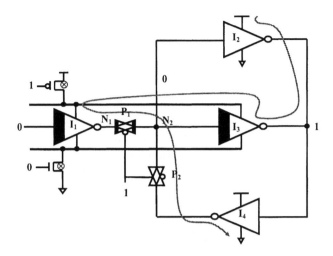

Figure 5.13. Leakage paths during standby mode due to reverse conduction sneak paths

suring that no more than one CMOS-MTCMOS gate pair has a common output node [9]. This will cut off the reverse conduction leakage path, which needs to flow from V_{dd} of one CMOS gate to the ground terminal of another CMOS gate.

Another way to eliminate the leakage path is to avoid sharing virtual V_{dd} and virtual *ground* lines for the MTCMOS gates, whose outputs are connected to the outputs of CMOS gates; that is, separate high-V_{th} sleep transistors for the MTCMOS gates are used. There is a drawback though, the local power switches of both polarities require more area, because sleep devices are not shared among multiple blocks such as in MTCMOS combinational logic. However, this will facilitate the sizing of the sleep transistors, and enhance the noise immunity because *ground* bounce on the virtual *ground* rails will be reduced. This is particularly advantageous for sensitive storage nodes.

Another way to cut off the sneak leakage paths is to design circuits with minimum connections between the CMOS and MTCMOS outputs. If an MTC-MOS block can be designed such that no output is connected through a leakage path to the output of a CMOS gate, then it is sufficient to connect all these MTCMOS gates to a common virtual *ground* or virtual V_{dd} line in order to eliminate leakage currents. In this case, a single polarity sleep device is sufficient to reduce the leakage currents by orders of magnitude.

In the next sections, interfacing MTCMOS and CMOS blocks, and how the value of high-V_{th} can impact the design of MTCMOS sequential circuits, will be briefly discussed.

5.8.   Interfacing MTCMOS and CMOS blocks

MTCMOS and CMOS logic families cannot be combined directly. Interfacing MTCMOS and CMOS blocks is also not straightforward. As mentioned in Chapter 4, during the standby mode, the MTCMOS gates can float because they are cut off from the V_{dd} and *ground* rails, and so they cannot directly drive the CMOS gates. If a CMOS gate has an intermediate floating input during the sleep mode, then very large short-circuit currents can develop. In order to eliminate these short-circuit currents, it must be ensured that the input signal is actively driven during the standby state. There are primarily two ways to accomplish this: (1) utilizing helper high-V_{th} gates, or (2) employing an intermediate latch circuit that holds the previous data during the entire standby period. Kao et al. introduced a leakage feedback gate which serves as an efficient interface between CMOS and MTCMOS logic. This will be illustrated later in Section 5.10.

5.9.   Impact of the High-V_{th} and Low-V_{th} values on MTCMOS Sequential Circuit Design

Impact of Parallel High-V_{th} Circuits

The flip-flops described so far employ parallel high-V_{th} inverters to maintain state during the standby mode. An example of such inverters is I_2 in the MTCMOS flip-flop in Figure 5.1, or I_4 in the configuration shown in Figure 5.14. These inverters improve the current driving capability. However, introducing parallel inverters to a critical path can degrade the overall performance, because the increase in capacitive loading can outweigh the benefit in the improved current drive.

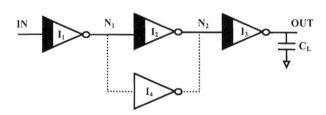

Figure 5.14.   Inverter chain showing insertion of possible high-V_{th} parallel device

This will be even more evident if a large difference exists between the high-V_{th} and low-V_{th}, and the circuit is designed so that the input rise/fall times are comparable to the output fall/rise times. In this case, the high-V_{th} parallel device turns *on* late and would not enhance the forward current drive. The degradation in speed outweighs any added current drive due to the parallel high-V_{th} device. Increasing the size of the high-V_{th} device is not helpful, because it increases the load capacitance on the critical path, and comes at the expense of extra area and increased leakage power. It would be more beneficial to simply size up the existing low-V_{th} inverter I_2 [10]. The difference between the high-V_{th} and low-V_{th} should be approximately 100mV [11].

MTCMOS Flip-Flop Design when a Large Difference Exists between High-V_{th} and Low-V_{th}

If the difference between the high-V_{th} and low-V_{th} is large, MTCMOS flip-flops must be devised to minimize the loading caused by the parallel high-V_{th}, but still being able to retain state during the standby mode. The flip flop structure is modified as in Figure 5.15 to minimize the loading on node N_3, whereas the low-V_{th} devices in the critical path are sized to maximize performance.

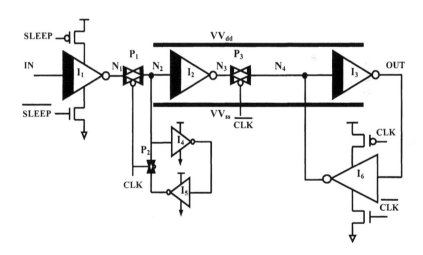

Figure 5.15. MTCMOS flip flop without parallel high-V_{th} paths [10]

This flip-flop design is also more immune to noise than a conventional MTCMOS flip flop, because the recirculation path is decoupled from node N_3. This benefit is even more important for the slave latch, because the output stage will then drive an outside block, which is completely decoupled from the recirculation path. Furthermore, the high-V_{th} feedback inverters can be decoupled from the MTCMOS critical path. This allows utilizing common virtual V_{dd} or virtual $ground$ lines that are shared among other MTCMOS flip-flops or logic circuits. As a result, the sneak leakage paths that exist when MTCMOS and CMOS outputs are connected are eliminated [10].

5.10.   Leakage Feedback Gates

Kao et al. introduced another way for state retention during the standby mode without using parallel high-V_{th} devices by devising leakage feedback gates [9].

These gates have the beneficial property of being able to actively drive their output to either V_{dd} or $ground$, and still be in a low leakage state during the standby mode. This is achieved regardless of the input signal and with smaller area overhead.

In standard combinational MTCMOS logic blocks, a single polarity sleep transistor is needed to completely eliminate the leakage currents. By utilizing only PMOS sleep transistors, it is possible for such an MTCMOS gate to hold a strong logic "0" during the standby mode. On the other hand, an MTCMOS gate that utilizes only NMOS sleep transistors can hold a strong logic "1" during the standby state. This is illustrated in Figure 5.16.

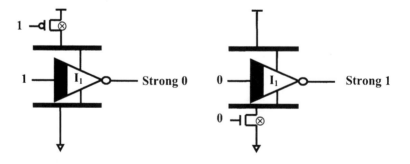

Figure 5.16.   Low leakage with driven outputs

The MTCMOS leakage feedback gate proposed by Kao et al. uses this concept, and holds either a strong logic "1" or a strong logic "0" during the standby state by selectively choosing either the PMOS or NMOS high-V_{th} sleep transistors to cut off leakage currents. Therefore, the output of the MTCMOS leakage feedback gate will always be driven to one rail or another, and will no longer float during the standby mode. Figure 5.17 shows the MTCMOS leakage feedback gate.

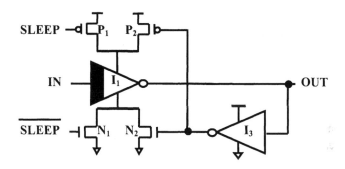

Figure 5.17. Leakage Feedback Gate

Standby mode: Sleep transistors P_1 and N_1 are turned off, thereby disconnecting the gate from the power supplies. Depending on the state of the latest output, one, but not both, helper sleep devices (P_2 or N_2) is turned on to retain the state of the output node. The high-V_{th} leaker devices P_2, N_2 and inverter I_3 are designed to be minimum size. This operation will take place independent on the state of the input signal during the standby mode.

The gate output is pulled up to V_{dd} by a leaking PMOS device, and pulled down to $ground$ by a leaking NMOS device. The difference in leakage current establishes the output voltage to the corresponding value. This is attributed to the imbalance of several orders of magnitude between the subthreshold leakage currents in high-V_{th} and those in low-V_{th} devices. This can be illustrated by examining the two scenarios in Figure 5.18.

Figure 5.19 presents two scenarios where an off high-V_{th} device is placed in series with an off low-V_{th} device for both holding conditions. The DC operating point for either of the two scenarios can be computed by equating the leakage currents of P_1 and N_1 as shown in Figure 5.18. The output voltage is fed back to the gate to set the proper leakage relationship between the pullup and pulldown paths, and hence, is called $leakage\ feedback$.

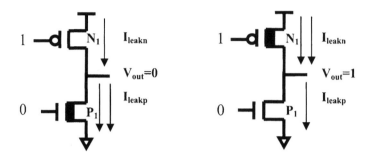

Figure 5.18.   Output state held by leakage currents.

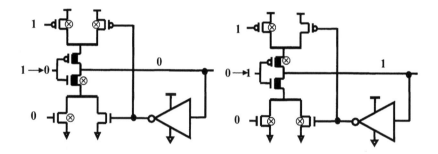

Figure 5.19.   Leakage feedback output retains state regardless of changes in input

In the first scenario of Figure 5.19, the input is driven *low* after entering the standby mode, and the output remains *low* , whereas in the second case, the input is driven *high* after the standby mode, and the output remains *high*. In the first case, even though a low-V_{th} PMOS is turned *on*, the path to V_{dd} must pass through strongly turned *off* high-V_{th} devices. On the other hand, the path from the output to *ground* is through a strongly turned *on* high-V_{th} device, and a turned *off* low-V_{th} device. Thus, the effective resistance to *ground* is much lower than the effective resistance to V_{dd}, so the output remains *low*.

In the second case, the opposite occurs: the low-V_{th} PMOS is turned *off*, but the high-V_{th} PMOS is turned *on*. Similarly, the resistance to V_{dd} is much lower than the resistance to *ground*, and the output is held *high*. Therefore, because of the leakage discrepancy between high-V_{th} and low-V_{th} devices, the DC operating point of the leakage feedback gate will always be close to V_{dd} or *ground* regardless of the input value.

Active mode: Sleep transistors P_1 and N_1 are turned *on*, thereby connecting the gate to the power supplies. Depending on the state of the output, either P_2 or N_2 will also be *on*, but the impact of these leaker devices on circuit performance is negligible. However, the addition of the minimum sized inverter I_3 causes a slightly higher loading, and the MTCMOS gate to be slower than a conventional gate. The savings in leakage power in the standby mode outweighs this slight increase in delay and dynamic power consumption in the active mode.

In the next section, Kao et al. illustrate how the leakage feedback gate is used in interfacing MTCMOS with CMOS blocks.

Interfacing MTCMOS and CMOS Circuits using Leakage Feedback Gates

As mentioned in Section 5.8, that MTCMOS and CMOS cannot be connected directly, because the output of the MTCMOS block floats when placed in the standby mode. If that MTCMOS block drives a CMOS stage, then the CMOS inputs will also float, causing severe short-circuit currents, noise-margin degradation, and ultimately gate malfunction. As a result, interfacing between MTCMOS and CMOS stages generally requires the use of extra latches or flip flops to maintain the signals during the sleep state in order to *actively* drive the CMOS block. This would result in significant overhead in extra circuitry and area that provides no other useful function except to ensure that the nodes do not float during the standby state. In addition, the extra circuitry can cause extra loading that degrades performance during the active period.

Leakage feedback gates can be used to interface between MTCMOS blocks and CMOS blocks. By modifying the last stage of the MTCMOS block to be a leakage feedback gate, the output signal never floats and can drive a CMOS block even during the standby mode.

To further streamline the leakage feedback gate, the feedback inverter can be collapsed into the next stage of logic, rather than utilizing a separate feedback inverter. For example, if an interface signal connects to only one CMOS inverting gate, then it is safe to simply use the output of this gate as the feedback signal to enable or disable the helper high-V_{th} sleep devices as portrayed in Figure 5.20.

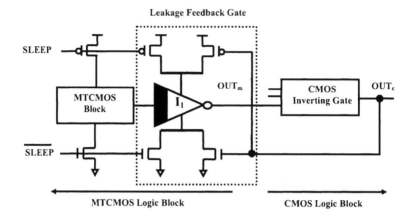

Figure 5.20.   Interface between CMOS MTCMOS

The output of the next stage CMOS gate produces the feedback signal because this will ensure that during the sleep stage, the output of the leakage feedback gate OUT_m will correspond to a voltage level that ensures the CMOS logic block output stays the same. For example, if OUT_m and the CMOS gate output OUT_c are opposite in polarity, then the feedback circuitry reinforces the leakage feedback gate to maintain that output level. However, if OUT_m and OUT_c are the same polarity, then the feedback signal causes the leakage feedback gate to flip the state of OUT_m.

It can be shown though that in this scenario, regardless of the state of OUT_m, the CMOS inverting gate would be driven to the proper output voltage simply due to the CMOS inputs to the gate.

Figure 5.20 shows the case where the MTCMOS output signal drives a single CMOS gate. If, though, this signal is routed to several different gates, then it would be necessary to use a standard leakage feedback gate like that of Figure 5.17, which uses a stand-alone feedback inverter to ensure that the CMOS block is driven with the same voltage levels during the sleep condition as before [10].

5.11.   Chapter Summary

This chapter presents different kinds of MTCMOS sequential circuits. For each circuit, the functionality and drawbacks are highlighted. The chapter discusses:

- The conventional MTCMOS latch circuit

- The MTCMOS balloon latch circuit

- The intermittent power supply scheme

- The auto-backgate-controlled MTCMOS scheme

- The virtual rails clamp circuit

- The leakage feedback gate

Leakage sneak paths in MTCMOS sequential circuits are also discussed, outlining their possible causes and solution to cut them off.

References

[1] S. Mutoh, S. Shigematsu, Y. Matsuya, H. Fukuda, T. Kaneko, and J. Yamada, "A 1-V Multithreshold-Voltage CMOS Digital Signal Processor for Mobile Phone Applications," *IEEE Journal of Solid-State Circuits*, vol. 31, no. 11, pp. 1795–1802, 1996.

[2] S. Shigematsu, S. Mutah, Y. Matsuya, Y. Tanabe, and J. Yamada, "A 1-V High-Speed MTCMOS Circuit Scheme for Power-Down Application Circuits," *IEEE Journal of Solid-State Circuits*, vol. 32, no. 6, pp. 861–869, 1997.

[3] H. Akamatsu, T. Iwata, H. Yamamoto, T. Hirata, H. Yamauchi, H. Kotani, and A. Matsuzawa, "A Low Power Data Holding Circuit with an Intermittent Power Supply scheme for sub-1V MT-CMOS LSIs," *Symposium on VLSI Circuits Digest of Technical Papers*, pp. 14–15, 1996.

[4] H. Akamatsu, T. Iwata, H. Yamauchi, H. Kotani, A. Matsuzawa, H. Yamamoto, and T. Hirata, "A Low Power Data Storage Circuit with an Intermittent Power Supply Scheme for Sub-1V MT-CMOS LSIs," *IEICE Transactions Electron.*, vol. E80-C, no. 12, pp. 1572–1577, December 1997.

[5] H. Makino, Y. Tujihashi, K. Nii, C. Morishima, Y. Hayakawa, T. Shimizu, and T. Arakawa, "An Auto-Backgate-Controlled MT-CMOS Circuit," *in Digest of Technical Papers, IEEE Symposium on VLSI Technology*, pp. 42–43, 1998.

[6] K. Nii, H. Makino, Y. Tujihashi, C. Morishima, Y. Hayakawa, H. Nunogami, T. Arakawa, and H. Hamano, "A Low Power SRAM using Auto-Backgate-Controlled MT-CMOS," in *Proceedings of the International Symposium on Low-Power Electronics and Design*, 1998, pp. 293–298.

[7] H. Notani, M. Koyama, R. Mano, H. Makino, and Y. Matsuda, "A 0.9-μA Standby Current DSP Core Using Improved ABC-MT-CMOS with Charge Pump Circuit," *Symposium on VLSI Circuits Digest of Technical Papers*, pp. 221–222, June 2001.

[8] K. Kumagai, H. Iwaki, H. Yoshida, H. Suzuki, T. Yamada, and S. Kurosawa, "A Novel Powering-Down Scheme for Low Vt CMOS Circuits," *Symposium on VLSI Circuits Digest of Technical Papers*, pp. 44–45, June 1998.

[9] J. Kao and A. Chandrakasan, "MTCMOS Sequential Circuits," *in Proceedings of European Solid-State Circuits Conference*, pp. 332–335, September 2001.

[10] J. Kao, *Subthreshold Leakage Control Techniques for Low Power Digital Circuits*, Ph.D. Thesis, Massachusetts Institute of Technology, May 2001.

[11] M. Hirabayashi, K. Nose, and T. Sakurai, "Design Methodology and Optimization Strategy for Dual-Vth Scheme using Commercially Available Tools," *in Proceedings of the International Symposium on Low Power Electronics and Design*, pp. 283–286, August 2001.

Further Reading

■ R. Ramanarayanan, N. Vijaykrishnan, and M. Irwin, "Characterizing Dynamic and Leakage Power Behavior in Flip-Flops," *in Proc. IEEE ASIC/SOC Conference*, pp.433-437, September 2002.

■ V. Zyuban and S. Kosonocky, "Low Power Integrated Scan-Retention Mechanism," *in Proc. IEEE International Symposium on Low Power Electronics and Design*, pp.98-102, August 2002.

■ J. Wang and H. Li, "0.9-V Sense-Amplifier-Based Reduced-Clock-Swing MTCMOS Flip-Flops," *in Proc. IEEE Asia-Pacific Conference on ASIC*, pp.271-274, August 2002.

■ M. Stan and M. Barcella, "MTCMOS with Outer Feedback (MTOF) Flip-Flops," *in Proc. IEEE International Symposium on Circuits and Systems*, Vol.5, pp.429-432, May 2003.

Chapter 6

MTCMOS DYNAMIC CIRCUITS

6.1. Introduction

In a digital CMOS circuit, reducing the supply voltage V_{dd} is the most effective approach to minimize dynamic power dissipation. This has been evident by the quadratic relation between the dynamic power and V_{dd} in Equation (2.1). However, reducing only the supply voltage seriously degrades the circuit's performance. One way to maintain performance is to scale down both V_{dd} and the threshold voltage V_{th}. However, reducing V_{th} increases the subthreshold leakage current exponentially (Equation (2.5)). Dynamic logic circuits such as Domino have a significantly worse tolerance to device subthreshold leakage compared to static CMOS circuits [1]. This makes it risky to utilize low threshold voltage (LVT) devices in order to improve the critical path delay [2]. Therefore, a trade-off exists between improving the gate's reliability (Noise Margin) and enhancing its speed.

A new Domino logic style, the High-Speed Domino (HS-Domino) is developed to resolve this speed-Noise Margin trade-off. The HS-Domino extends the Domino's operation into the Deep Sub-Micron (DSM) regime with no degradation to the gate's Noise Margin (NM). A new Multi-Threshold (MTCMOS) scheme for dynamic logic styles is then presented. This scheme is applied to both HS-Domino and Domino Differential Cascode Voltage Switch (DDCVS) logic. The MTCMOS implementation substantially reduces the sub-threshold leakage current during the standby mode, while attaining high performance and low dynamic power values during the active mode.

6.2.   Clock-Delayed Domino Logic: Overview

Figure 6.1 shows an 8-input Clock Delayed Domino (CD-Domino)[1] OR gate. It is a non-inverting structure and consists of a dynamic gate stage which implements the function in the NMOS logic block, a static CMOS inverter, which provides the circuit's output, and a PMOS keeper transistor which restores the logic at the Domino output node.

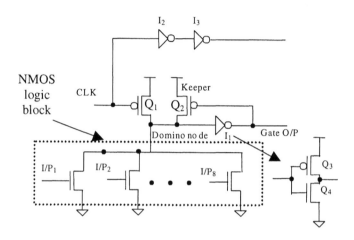

Figure 6.1.   8-input Clock-Delay Domino OR gate

The conventional Domino operates as follows: during the precharge phase (when the clock is low), the Domino output is precharged to V_{dd} and the keeper is turned *on*. When the clock goes *high* (the evaluation phase), depending on the inputs, the Domino output is either discharged to *ground* or remains *high* at V_{dd}. If all the inputs are low, the keeper will keep the output high. In a cascaded set of logic blocks, each state evaluates and causes the next stage to evaluate in the same way a line of dominos fall, hence the name, Domino. Any number of logic stages can be cascaded, provided that the sequence can evaluate within the evaluate clock phase.

An NMOS footer transistor is usually added in series between the NMOS logic block and *ground* to guarantee that the precharge and evaluation never take place simultaneously. In CD-Domino, the NMOS footer transistor is eliminated, and self-timed delays are used for the precharge and evaluation clocks. This delay is achieved through a delay element, which is usually represented as two cascade inverters (I_2 and I_3 in Figure 6.1). The self-timed clock output of the delay element tells the next gate when the data output is ready, as represented in Figure 6.3.

Therefore, for a pipeline of gates implemented in CD-Domino, the clock supplying the downstream gates is delayed with the flow of data so that the NMOS foot switches can be removed to achieve higher performance. Figure 6.2 shows a domino pipe stage in a two-phase clocking methodology. The pipe stage is composed of a logic depth of four domino gates and four inverters producing an overall depth of eight gate delays. There are three very important timing constraints that need to be satisfied for the correct functionality of CD-Domino circuits.

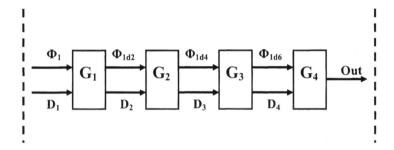

Figure 6.2. ϕ_1 pipeline stage with four levels

(1) The stage clock delay (the delay of inverters I_2 and I_3 together) must be matched to the evaluate domino gate delay (the delay of the pulldown logic block plus the delay of output inverter I_1). This guarantees that the evaluation clock signal and the gate input signal will arrive in alignment.

(2) The stage clock delay (the delay of I_2 and I_3 together) must be matched to the precharge domino gate delay (the delay of the PMOS precharge device plus the delay of output inverter I_1). This guarantees that the precharge clock signal will arrive in conjunction with the input zero signal. As a result, this timing constraint eliminates contention between the PMOS precharge device trying to pull up the domino node to V_{dd}, and the NMOS pulldown logic block attempting to pull down the same domino node to $ground$. Thus, short circuit currents are eliminated.

(3) An NMOS footer must be retained in the first stage of the pipeline. This is necessary because the Phase 1 logic block will most likely be driven by a Phase 2 logic block, but the output from the Phase 2 block can become valid when the first stage of the Phase 1 block is still precharging which causes contention and consequently short-circuit currents.

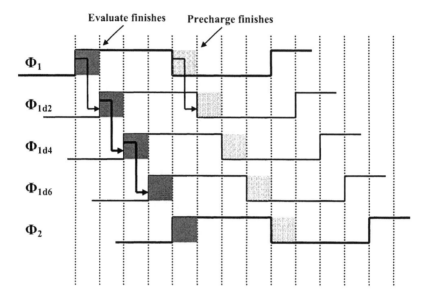

Figure 6.3.   Clocking methodology for CD-Domino showing evaluation and precharge times

One of the drawbacks of this clock delayed methodology is that tighter constraints are applied to the precharge time. In a conventional domino, all the gates begin the precharge cycle at the same time, and thus, can take the entire half cycle to precharge. In the CD-domino case, though, both the evaluation and precharge of each gate must be completed in the allotted *two inverter delay slot*, because the precharge of each gate proceeds serially.

The Speed - Noise Margin Trade-Off in Domino Circuits

Whether the domino circuit is implemented in the conventional scheme (with an NMOS footer) or in the clock-delayed scheme (footless), the domino gate is vulnerable to a speed-noise immunity trade-off. This can be explained as follows: During the **precharge** phase, the Domino node is charged to "1", and the keeper transistor Q_2 turns *on* to maintain the voltage of the Domino node. When the clock CLK goes *high* (the evaluation mode), depending on the inputs, the Domino node is either discharged to *ground* or remains *high*.

At the beginning of the **evaluation**, the keeper is *on*, charging the Domino node to "1". At the same time, the pulldown devices are trying to discharge the Domino node. This is called contention, where one device is trying to charge a node while another device is trying to discharge it. Contention slows down the evaluation, and increases the dynamic power dissipation because of the large current flowing from V_{dd} to *ground* during the evaluation. Therefore, it is preferable to size down the keeper to reduce the contention current flowing through Q_2, and thus enhancing the evaluation speed. A small keeper size, however, cannot compensate for any leakage currents or charge sharing when all inputs are "0s" during the evaluation phase. This ultimately degrades the gate's NM. This is the basic speed-NM trade-off in conventional CD-Domino circuits, and the effect is even more severe at lower V_{th}'s as the technology scales down. The contention period is plotted in Figure 6.4, where the Domino node starts to switch while the keeper is still *on*.

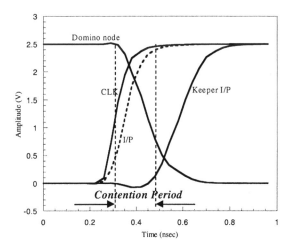

Figure 6.4. Waveforms of CD-Domino(V_{dd}=2.5V)

Figure 6.5 shows the NM of an 8-input conventional Domino OR gate for different V_{th} values, while keeping the ratio W_{keeper}/W_n constant and equal to 1/10, where W_{keeper} and W_n are the widths of the PMOS keeper and NMOS pulldown devices respectively. The 8-input OR gate has been chosen, because Domino logic is usually used for wide fanin OR gates. OR gates experience the worst case leakage current when all the inputs are "0s".

It is clear in Figure 6.5 that the NM drops by approximately 1mV for every 1mV decrease in V_{th}. The NM is defined as the input voltage above *ground* that causes a 10% drop off V_{dd} at the Domino node [1]. The NM is set to 10%

of V_{dd} in this work .

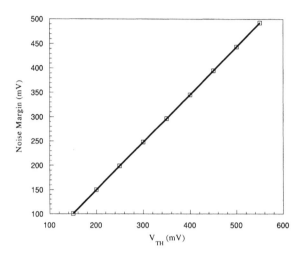

Figure 6.5.   Noise Margin of CD-Domino vs. V_{th}

Figure 6.6 illustrates the normalized delay of a 3-stage chain of 8-input CD-Domino OR gates with a fanout of three versus V_{th} for two cases: (1) constant W_{keeper}/W_n ratio (i.e., ignoring the NM), and (2) a controlled NM (at least 10% of V_{dd} by increasing the W_{keeper}/W_n ratio). The constant W_{keeper}/W_n curve shows that the performance of the CD-Domino increases as V_{th} decreases. However, this curve neglects the NM which leads to an impractical design. On the other hand, to control the NM, the keeper has to be sized up, leading to a greater contention current at the beginning of the evaluation. This degrades the gate's speed particularly at low V_{th}s and ultimately increases the delay (the controlled NM curve).

Therefore, Domino circuits are not suitable for DSM technologies because of their high leakage currents and degraded performance [3]. The proposed HS-Domino style resolves this speed-NM trade-off problem.

6.3.   HS-Domino Logic

The architecture of the 8-input HS-Domino [4] [5] OR gate is shown in Figure 6.7. It is similar to the CD-Domino except that the gate output is connected to the keeper through an NMOS (N_1) and a PMOS (P_1) device, whose gates are connected to the delayed clock signal.

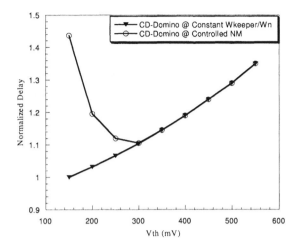

Figure 6.6. Delay of CD-Domino vs.V_{th}

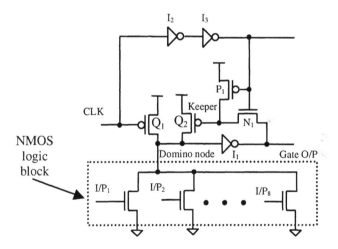

Figure 6.7. An 8-input HS-Domino OR gate

The HS-Domino operates as follows: when the clock is *low* during precharge, the Domino node is precharged to V_{dd}, transistor N_1 is *off*, P_1 is *on*, charging the gate of the keeper transistor Q_2 to V_{dd}, thus turning Q_2 *off*. Q_2 is, therefore, *off* at the beginning of the evaluation phase. As a result, contention is eliminated between the keeper and the pulldown devices during the evaluation. Therefore, the Domino gate evaluates faster, and no contention current exists.

When the delayed clock becomes "1", and if the Domino node evaluates to "0", the gate output is "1", and N_1 is *on*, thus keeping Q_2 *off*. However, if all the pull-down devices are *off*, the Domino node stays at "1", causing the gate output to be "0", which in turn, discharges the keeper's gate through N_1. Therefore, the keeper turns *on* to maintain the voltage of the Domino node at V_{dd} and to compensate for any leakage currents. Consequently, the HS-Domino solves the contention problem by turning the keeper *off* at the start of the evaluation cycle. Therefore, the keeper width can now be sized up as V_{th} scales down to maintain a controlled NM without increasing the contention, and speed degradation.

The contention-free operation is clearly demonstrated in Figure 6.8, where the Domino node starts switching, while the keeper's input (gate) remains *high*. The small *dip* to the keeper's input is far from sufficient to start turning the keeper *on*. It should also be noted that the CLK signal must arrive before the input, to prevent contention between the PMOS precharge device Q_1 and the pulldown devices. This is an important timing constraint in CD-Domino circuits in general [6].

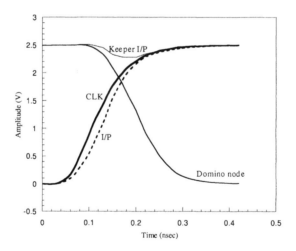

Figure 6.8.   Waveforms of HS-Domino gate (V_{dd}=2.5V)

Speed and Power Comparison

In order to compare the performance of the HS-Domino with that of the CD-Domino, a 3-stage chain of 8-input OR gates with a fanout of three, is simulated in 0.25 μm CMOS technology at V_{dd}=2.5V. The normalized delay versus V_{th} is presented in Figure 6.9. The delay curves of the CD-Domino with a controlled NM and constant W_{keeper}/W_n ratio in Figure 6.6 are re-plotted to illustrate the speed advantage of the HS-Domino. Figure 6.9 indicates how the delay of the HS-Domino circuit continues to decrease as V_{th} is scaled down without degrading the NM. A slight speed difference begins to develop between the HS-Domino and the CD-Domino at the constant W_{keeper}/W_n ratio, as V_{th} decreases. This occurs because the loading at the Domino node increases as the keeper is sized up to keep the NM intact. The HS-Domino has a 30% speed advantage over the CD-Domino at very low V_{th}s, while controlling the NM.

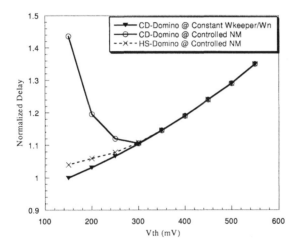

Figure 6.9. Normalized delay vs V_{th}

Figure 6.10 compares the HS-Domino with the CD-Domino in terms of the dynamic power at a clock frequency of 500MHz with the same OR gate chain. The HS-Domino with a controlled NM reduces power dissipation by 15% to 24% compared to that of the CD-Domino at a controlled NM. This is attributed to: (1) the elimination of contention, indicating that there are no short-circuit currents flowing from V_{dd} to $ground$, and (2) N_1 and P_1, are minimum size devices which have a minor effect on the total clock loading (3% of the total clock loading), and consequently, the overall power dissipation. The major loading is attributed to the precharge device and the clock delay inverters of each domino gate. At very low V_{th} values, the power of the HS-Domino increases, because the keeper is sized up to maintain the NM which increases the loading at the Domino node.

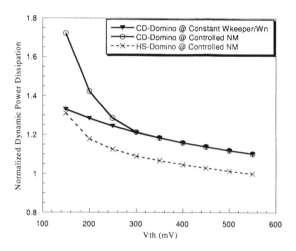

Figure 6.10. Normalized dynamic power vs V_{th}

Nevertheless, the HS-Domino proves to consume less dynamic power than that of the conventional case. With these delay and power savings, the HS-Domino offers an energy saving of up to 60% at very low V_{th}s, while keeping the NM intact. It should be mentioned that the superiority of the HS-Domino would be even more evident in DSM technologies, where leakage contributes heavily to the design.

Figures 6.11 and 6.12 show the layout of a 3-input OR gate, implemented by using the HS-Domino and CD-Domino. The HS-Domino layout has a minimal area overhead compared to that pf the CD-Domino, because N_1 and P_1 are minimum size devices.

Adder Implementations

To further illustrate the superiority of the HS-Domino over the conventional CD-Domino, a 4-bit Ripple Carry (RC) adder, and a 4-bit Carry Look Ahead (CLA) adder are used as test vehicles. Both adders are implemented in $0.18\mu m$ CMOS technology with V_{dd}=1.8V, high-V_{th}=500 mV, and CLK=500MHz. Table 6.1 summarizes the percentage savings that the HS-Domino provides over the CD-Domino while the NM is kept intact.

Both adders implemented in HS-Domino offer significant delay, power, and energy savings compared to the CD-Domino. It should be noted that both adders achieve less savings than the 8-input OR gate case. This can be ex-

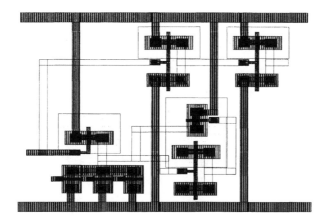

Figure 6.11. Layout of CD-Domino

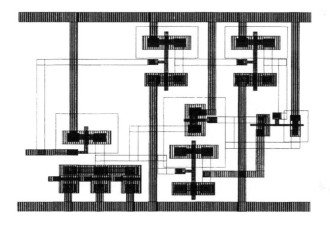

Figure 6.12. Layout of HS-Domino

plained; both adders are composed of some gates that have stacked NMOS devices such as the AND and XOR gates. Stacked devices reduce leakage due to the de-biasing effect which can make V_{gs} negative for the top NMOS in the stack [7] which exponentially reduces the leakage (as illustrated in Chapter 2).

The leakage power of the HS-Domino is also calculated, and is equal to that of the CD-Domino (the constant W_{keeper}/W_n and controlled NM). For the 0.18μm CMOS technology, the leakage current increases by an order of magnitude for every 85mV reduction in V_{th}.

One of the major concerns in the design of the conventional CD-Domino, as well as the proposed HS-Domino, is the timing constraint. The input signals

Table 6.1.   Percentage savings in adders with controlled NM

V_{th} (mV)	RC Adder (% Saving)			CLA Adder (% Saving)		
	Delay	Power	Energy	Delay	Power	Energy
500	1.20	8.20	9.40	1.70	9.35	10.9
450	1.20	9.00	10.0	1.80	9.50	11.0
400	1.20	9.30	10.1	2.00	9.60	11.1
350	1.20	9.30	10.1	2.25	9.70	11.3
300	1.30	9.40	10.3	2.50	10.0	11.5
250	1.30	9.70	10.9	2.70	10.2	12.6
200	4.20	10.9	14.6	4.60	10.4	14.5
150	11.5	11.5	21.6	13.9	12.3	24.5
100	21.5	15.0	33.3	28.4	17.2	39.2
50	33.2	21.0	47.7	45.0	23.9	57.5

should arrive after the arrival of the clock signal to achieve the declared energy savings. Therefore, the devices of the domino gate, and the clock delaying inverters are properly sized to fulfill such a timing constraint. This is applied in the implementation of the 3-stage chain of 8-input OR gates, as well as the 4-bit RC and CLA adders. However, such timing constraints would become a challenge in more complicated domino networks.

It is evident that the HS-Domino resolves the classic speed-NM trade-off in conventional domino circuits, but the leakage power is still large. Therefore, a Multi-V_{th} (MTCMOS) implementation of the HS-Domino (MHS-Domino) is devised which achieves low values of leakage power in the standby mode, while attaining the high performance and low power of single V_{th} HS-Domino circuits during the active mode.

The following two sections explain the MTCMOS implementation of the HS-Domino, and its superiority over the conventional MTCMOS implementation of CD-Domino logic.

6.4.   MTCMOS CD-Domino Logic: Analysis and Overview

The Multi-V_{th} implementation of the domino gates utilize both high-V_{th} (HVT) and low-V_{th} (LVT) transistors that allow the overall circuit to achieve extremely low leakage currents in the standby mode, yet do not reduce the performance in the active mode.

In order to determine which transistors should have HVTs or LVTs, and at which phase of operation standby mode should occur, four phases are investigated in domino logic: (1) the evaluation phase, (2) the precharge phase, (3) standby at the evaluation phase, and (4) standby at the precharge phase.

Devices in the critical charge/discharge path should always be assigned a LVT so as not to degrade the gate's speed. On the other hand, devices that switch during the *delay uncritical* precharge mode should be HVT to dissipate the minimum leakage power during standby. According to Figure 6.1, Table 6.2 is constructed and summarizes which transistors should have HVTs or LVTs, according to the four modes of operation previously mentioned. N_{I2} and P_{I2} are the N and P devices of I_2, whereas N_{I3} and P_{I3} are the N and P devices of I_3.

Table 6.2. Type of Transistors in the Domino Logic ("H" symbolizes a HVT device, "L" a LVT device, and "X" a *don't care state* device (i.e., can be either a HVT or LVT))

Mode	Q_1	Q_2	Q_3	Q_4	N_{I2}	P_{I2}	N_{I3}	P_{I3}
Precharge	X	X	X	X	X	X	X	X
Standby at Precharge	X	X	H	X	H	X	X	H
Standby at Evaluation	H	H	X	H	X	H	H	X
Evaluation	X	X	L	X	L	X	X	L

Table 6.2 can be further divided into two tables, either standby at precharge (Table 6.3) or standby at evaluation (Table 6.4).

Table 6.3. Standby at Precharge : **Rejected**

Mode	Q_1	Q_2	Q_3	Q_4	N_{I2}	P_{I2}	N_{I3}	P_{I3}
Precharge	X	X	X	X	X	X	X	X
Standby	X	X	**H**	X	**H**	X	X	**H**
Evaluation	X	X	L	X	L	X	X	L
Device kind	X	X	—	X	—	X	X	—

As indicated in Table 6.3, transistors Q_3, N_{I2}, and P_{I3} should be HVT during the standby mode. However, these same devices should also be LVT during the evaluation mode. This discrepancy indicates that the standby mode cannot

Table 6.4.   Standby at Evaluation : **Correct**

Mode	Q_1	Q_2	Q_3	Q_4	N_{I2}	P_{I2}	N_{I3}	P_{I3}
Precharge	X	X	X	X	X	X	X	X
Standby	H	H	X	H	X	H	H	X
Evaluation	X	X	L	X	L	X	X	L
Device kind	H	H	L	H	L	H	H	L

take place at the precharge phase. Devices Q_3, N_{I2} and P_{I3} that are causing the discrepancy are highlighted in bold in Table 6.3.

On the other hand, Table 6.4 shows no contradiction in the kind of device during any mode of operation. Thus, the standby mode is chosen to exist during the evaluation phase. In compliance with Table 6.4, an MTCMOS CD-Domino implementation is devised [8] [9]. Figure 6.13 illustrates the schematic of this MTCMOS CD-Domino gate, where the LVT devices are highlighted.

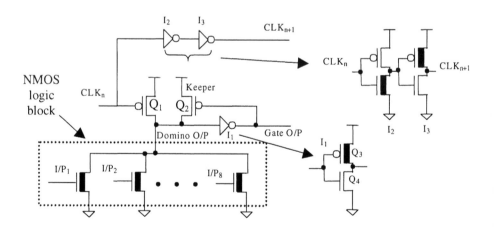

Figure 6.13.   An 8-input MTCMOS CD-Domino OR gate

Evaluation Mode

Before the domino gate enters the evaluation mode, the domino output node is precharged *high*, while the inputs, CLK_n, CLK_{n+1} are all *low*. During evaluation, the CLK_n goes *high* and data arrives to the inputs, the domino will evaluate through the fast LVT NMOS pulldown devices and LVT PMOS of I_1. Furthermore, the CLK_n ripples quickly through the LVT NMOS of I_2 and PMOS of I_3 to supply the clocking signal to the next level of domino logic. It is very important that the delay of both I_2 and I_3 is matched to the delay of the domino gate (pulldown devices and I_1) such that the next data arrival is correctly timed with the next evaluation clock. The threshold voltage assignments are in agreement with Table 6.4.

Precharge Mode

During the precharge phase, the charging and discharging path pass through the HVT devices in agreement with Table 6.4. In order to eliminate contention (explained in the CD-domino timing) in the next domino stage, the delay of both I_2 and I_3 is matched to the delay of the precharge devices P_1 and I_1. Because the HVT devices perform the precharge functions, the precharge time is longer than for the case where all the LVT devices are employed. As a result, the CLK pulse width increases as it travels downstream to the gates in order to align the precharge transition with the CLK. Since the precharge time is not in the critical path, the use of HVT devices is acceptable.

Standby Mode

During standby, the gate operates in the evaluation mode (CLK=$high$) in order to shut off the HVT devices. The HVT transistors Q_1, Q_2, Q_4, the PMOS of I_2, and the NMOS of I_3 are off, reducing the leakage current during the standby phase. Furthermore, the primary inputs are forced $high^2$ to ensure that the domino output node completely evaluates and remains in a strong "0" state. Otherwise, the internal node can float, introducing short-circuit currents. Ensuring that the gate output has a strong "1" guarantees that all the subsequent gates in the pipeline will evaluate in a cascaded fashion. The resultant datapath will thus be in a low leakage state, where all the HVT devices are strongly turned off.

Timing Constraints in MTCMOS Domino Logic

As mentioned in Section 6.2, in a CD-Domino logic implemented in a single-V_{th}, both the evaluation and precharge for each domino stage must be completed in the allocated two inverter delay slot. In a pipeline of domino logic, gate N+1 will begin to precharge only after gate N has already precharged, producing zero outputs. However, in the dual-V_{th} CD-Domino design, the inverter delays during the evaluation and precharge phases, are not identical. This allows more time for the precharge to take place during the clock cycle. It should also be mentioned that the timing and design constraints for a

single-V_{th} CD-Domino pipelines, stated in Section 6.2, must also be applied to
MTCMOS CD-Domino designs.

The operation of the dual-V_{th} CD-Domino is more complicated than the
single-V_{th} case, because the precharge delay-noncritical transitions are slow,
whereas the evaluation delay-critical transitions are fast. This occurs because
the precharge delay-noncritical transitions pass through the high-V_{th} devices,
while the evaluation transitions flow through the low-V_{th} devices. This results in a stretching of the clock pulses in order to attain the matching of both
the evaluation and precharge clock signals with the data propagation. The
timing diagram of the two-phase dual-V_{th} CD-Domino pipeline is shown in
Figure 6.15 [9].

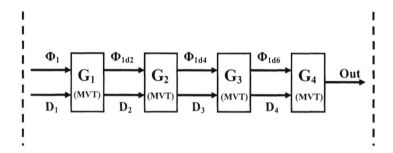

Figure 6.14. ϕ_1 pipeline stage with four levels

It can be seen in Figure 6.15 that the precharge clock signal (the falling edge
of the clock) for the downstream gates continues to be slowed down through
the high-V_{th} devices. On the other hand, the evaluation clock signal (the rising edge of the clock) for the downstream gates is unchanged. It is assumed
that the stage delay of the evaluation clock signal is annotated by $T_{evaluate}$,
where $T_{evaluate}$ = the stage evaluation delay through the low-V_{th} devices of
the clock inverters. Similarly, the stage delay of the precharge clock signal is
annotated by $T_{precharge}$, where $T_{precharge}$ = the stage precharge delay through
the high-V_{th} devices of the clock inverters. Thus, for each stage, the rising edge
is delayed by $T_{evaluate}$, while the falling edge is delayed by $T_{precharge}$. This
results in the clock pulse being stretched by $T_{precharge}$-$T_{evaluate}$ per stage. For
the two-phase domino pipeline with a 50% duty cycle and four logic stages per
phase, represented in Figure 6.15, the time available to perform a precharge is
$T_{precharge}=7/4T_{evaluate}$ [8]. For the case where the duty cycle is not 50%, then
more time can be allotted for the precharge phase. In general, if the total evaluation clock period (the high pulse width) is $N \times T_{evaluate}$, and the precharge

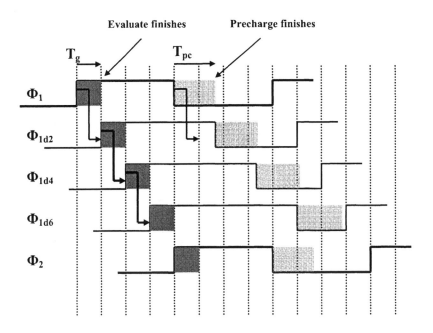

Figure 6.15. Clocking methodology for MTCMOS CD-Domino showing evaluate and precharge times

clock period (the low pulse width) is $M \times T_{evaluate}$ (so the total clock period is $(N+M) \times T_{evaluate}$), then the time available for precharge per stage is

$$T_{precharge} = \frac{(M + N - 1)T_{evaluate}}{N} \qquad (6.1)$$

Problems with the Dual-V_{th} Domino Logic

There are two main problems that are associated with the dual-V_{th} domino logic proposed in [8] [9].

- Forcing the inputs to be $high$ requires a gating circuitry at each input in the primary stage. This not only increases the dynamic and leakage power in the active and standby modes, respectively, but also increases the delay and area of the total circuit.

- The dual-V_{th} domino implementation does not take the noise margin into consideration, and also has contention at the beginning of the evaluation phase between the keeper and the pulldown devices as explained in Section 6.2.

In order to resolve the speed-NM trade-off, and thus eliminate the contention, as well as achieving ultra low leakage values with minimal area overhead, an MTCMOS version of the HS-Domino (presented in Section 6.3) circuit is proposed.

6.5.   MTCMOS HS-Domino (MHS-Domino) Logic

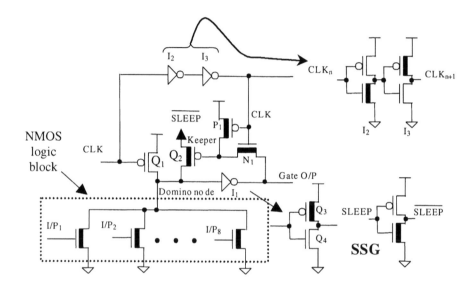

Figure 6.16. An 8-input MHS-Domino OR Gate

The MHS-Domino gate, illustrated in Figure 6.16, is similar to the HS-Domino gate, except that the keeper's source is connected to a $\overline{SLEEP}$ signal instead of V_{dd}. The MHS-Domino circuit also employs a Sleep Signal Generator (SSG), which is realized by an inverter as shown in the schematic. In this design, the $SLEEP$ signal is "0" for active operation and "1" for standby [4] [5].

During the active mode ($\overline{SLEEP}$ is "1"), the operation of the MHS-Domino gate is the same as that of the HS-Domino. Therefore, the keeper's source is connected to V_{dd} through the SSG block. In this scheme, all the transistors involved in the evaluation are chosen to be LVT devices to reduce the delay. These devices are the NMOS logic block, Q_2, Q_3, P_1, N_1, the NMOS transistor of I_2, and the PMOS transistor of I_3. The LVT devices are highlighted in

Figure 6.16.

In the standby mode, $\overline{SLEEP}$ becomes "0", and the clock is $high$. The Domino node has two possible values: "1" or "0". If the node is "0", the gate output is "1'," and therefore the input to the following gate is "1". When the Domino node is "1", the output is "0", and N_1 is on which turns on the keeper. Since, $\overline{SLEEP}$ is "0", the Domino node is discharged through the keeper, causing the gate output to switch to "1". This transition is fast because N_1 and the keeper are LVT. Consequently, at the beginning of the standby mode, all the MHS-Domino gates change their output to "1" regardless of the input values. Therefore, input circuitries are not needed which reduces the hardware and power. During the standby mode, all the HVT transistors Q_1, Q_4, the PMOS of I_2, the NMOS I_3, and the PMOS of the SSG turn off to reduce the leakage current. It should be also noted that some short circuit currents can flow in inverter I_1, if the Domino node is initially $high$ when the domino gate enters the standby mode. To eliminate this problem, primary inputs for the first gate will have to be forced $high$.

Speed and Power Comparison

The normalized delay of a 3-stage chain of 8-input Domino OR gates with a fanout of three for the MHS-Domino circuit versus V_{th} is visualized in Figure 6.17. The delay curves of the conventional MTCMOS CD-Domino [9] with a controlled NM and constant W_{keeper}/W_n ratio are also plotted on the same graph to illustrate the speed advantage of the MHS-Domino. Simulations are conducted in 0.25 μm CMOS technology, V_{dd}=2.5V and CLK=500MHz. Figure 6.17 indicates that the delay of the MHS-Domino circuit at the controlled NM continues to decrease as V_{th} is scaled down, because there is no contention current.

Figure 6.18 compares the MHS-Domino circuit with the MTCMOS CD-Domino circuit in terms of the dynamic power at CLK=500MHz. Although the MHS-Domino circuit introduces a slightly higher loading and an SSG, it actually has a significantly lower power dissipation compared to that of the conventional MTCMOS domino design. This is attributed to the following: (1) the elimination of contention in the MHS-Domino gate which means that there are no short-circuit currents during switching, (2) N_1 and P_1 are minimum sized devices, contributing to a very small loading effect, and (3) the sleep signal generator consumes very little dynamic power, because its output is always "1" during the active mode, and "0" during standby. Thus, no switching occurs except during the transition from one mode to the other. These transitions are

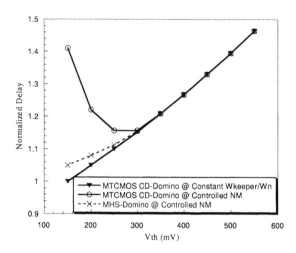

Figure 6.17. Normalized delay vs V_{th}

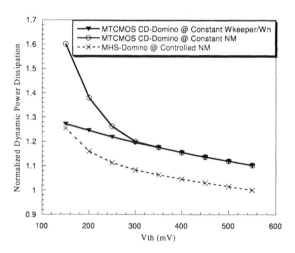

Figure 6.18. Normalized dynamic power vs V_{th}

not frequent, and their power dissipation penalty becomes less significant if the system stays in the idle state most of the time.

A comparison between the normalized leakage power of the conventional MTCMOS CD-Domino and MHS-Domino is also shown in Figure 6.19. The MHS-Domino consumes slightly higher leakage power than the MTCMOS CD-Domino (a controlled NM). However, for low-V_{th} values, the MHS-Domino consumes less leakage power. The difference in leakage power between the two cases is negligible, since the leakage current is in the order of pico-amperes. In Figures 6.17, 6.18 and 6.19, the HVT is taken to be 550mV, while the LVT

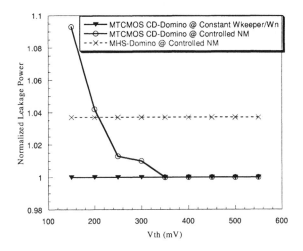

Figure 6.19. Normalized leakage power vs V_{th}

is varied and is denoted by V_{th}.

Therefore, the MHS-Domino eliminates contention, enhances performance, reduces dynamic power dissipation, and exhibits very low leakage current during standby with minimal area overhead. In addition, it should be mentioned that a single SLEEP generator is used for each domino block. This may cause routing overhead for large domino circuits.

The scheme that is used to convert a HS-Domino to an MTCMOS logic style is generic, and may be applied to other dynamic logic styles. Although the speed-NM problem is solved in that Domino Dual Cascode Voltage Switch Logic (DDCVSL), the new MTCMOS scheme is extended to a DDCVSL (MDDCVSL) to reduce the leakage power. A short overview of DDCVSL is first presented followed by the proposed MDDCVSL design.

6.6.   Domino Dual Cascode Voltage Switch Logic (DDCVSL)

Figure 6.20 presents a 2-input XOR gate implemented in DDCVSL [10]. It is clear that during the precharge phase (CLK=0), both keeper transistors Q_1 and Q_2 are off. Unlike Domino logic, the keeper transistors are off at the start of the evaluation phase (CLK=1) which reduces the power and delay caused by the contention.

One branch implements the required function, while the other branch im-

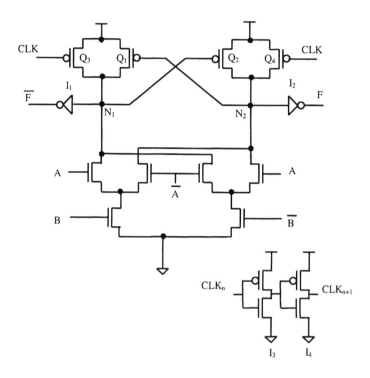

Figure 6.20. A two input DDCVS XOR logic gate

plements its inverse. DDCVSL is considered a general purpose logic style, because it can be used to implement inverting and non inverting logic circuits. In addition, DDCVSL has a higher area efficiency in implementing complex logic gates. Most of the complex logic functions can be implemented in a single gate, attributing to DDCVSL's speed over CMOS circuits. Thus, DDCVSL is efficient in implementing gates such as arithmetic circuits and multiplexer MUX style logic gates with XOR functionality.

MDDCVS Implementation

The MDDCVS has a similar implementation to the conventional double-branched DDCVS, except for a sleep signal generator (SSG), which connects the sources of the PMOS keeper devices (Q_1 and Q_2) to V_{dd} during the active mode (the conventional DDCVSL), or to *ground* during the sleep mode. Figure 6.21 shows a typical clock-delayed 2-input XOR MDDCVS stage.

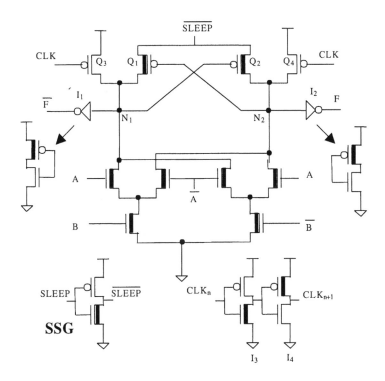

Figure 6.21. A two input MDDCVS XOR logic gate

During normal operation, the $\overline{SLEEP}$ signal is $high$ and the MDDCVS circuit operates exactly like a conventional DDCVS circuit. Because the precharge time is not critical, the transistors involved in the precharge process should be HVT to reduce leakage during standby. These transistors are Q_3, Q_4, the NMOS of inverters I_1 and I_2, the PMOS of I_3, the NMOS of I_4 and the PMOS of the SSG; when the CLK is $high$ (when the logic is evaluated), the transistors involved in the evaluation should be LVT to speed up the logic gate. The transistors that switch in the evaluation phase are the NMOS block transistors, the NMOS of I_3, the PMOS of I_4 and the PMOS transistors of I_1 and I_2. The LVT devices are highlighted in Figure 6.21.

During the standby mode, both the CLK and SLEEP signal are $high$, turning off the HVT devices. Whether or not standby occurs right after the precharge or the evaluation phase is not an issue. In both cases, whatever the input values to the gate are, a state will always be reached where one branch is on (a path to $ground$ exists) and the other is off. To illustrate this: assume that the value of the inputs to the N-block cause node N_1 to be low, and thus, N_2 to be $high$. Q_2 is, therefore, turned on, allowing node N_2 to start discharging through Q_2 until N_2 eventually reaches "0". If the inputs were to cause N_2 to be low and N_1 to be $high$, Q_1 will turn on allowing node N_1 to start discharging through Q_1, until N_1 eventually reaches "0".

Therefore, regardless of the inputs to the gate, both N_1 and N_2 will be "0" during the standby mode. The time taken to reach "0" for both N_1 and N_2 is calculated to be approximately 200psec. The result is that both nodes F and $\overline{F}$ will go $high$, which causes the input NMOS devices in the successive stage to turn on completely, and pulldown the two internal nodes (N_1 and N_2 of the next stage) to $ground$ very quickly. In the second stage, the discharging time takes approximately 60psec. This can be generalized to any other cascaded gates in the pipeline. Therefore, the first stage usually takes the longest time to reach the standby state, whereas the other consecutive gates in the pipeline take shorter periods of time. This is not critical, especially in burst-mode systems, where over 95% of the time the system is idle. The 200psec is, by far, negligible compared to the period of time (seconds or even minutes) a burst-mode system can be idle for.

An important advantage of the MDDCVS is that it does not require specific input values to the gate at the standby mode. This eliminates any increase in area, power, or delay, as a result of the gating circuitry for the inputs to the DDCVS gate.

Speed and Power Comparison

To verify the functionality and benefit of the MDDCVS, simulations are performed on a pipeline of three MDDCVS XOR gates with a fanout of three operating at 500MHz, and using a $0.25\mu m$ CMOS technology at a supply voltage of 2.5V. XOR gates are used as a test vehicle, because DDCVS logic is normally used to implement XOR and MUX circuits due to its differential compact nature. Figure 6.22 depicts the normalized delay of the 3-stage chain of the DDCVS gates versus V_{th} for three cases: (1) single V_{th} DDCVS at constant W_{keeper} (ignoring the NM), (2) single V_{th} DDCVS at a controlled NM, and (3) MDDCVS at a controlled NM. W_{keeper} is the size of the keeper transistors Q_1 and Q_2.

Similar to the Domino logic case, the NM is defined as the input voltage above $ground$ that causes a 10% drop from V_{dd} at the DDCVS internal nodes

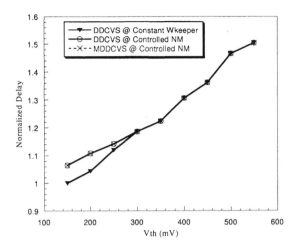

Figure 6.22. Normalized delay vs V_{th}

(N_1 or N_2). The NM is set to 10% of V_{dd}. Figure 6.22 shows that the MD-DCVS in the controlled NM has similar performance to the single V_{th} at the controlled NM case. The constant keeper curve represents the maximum possible gain in speed by lowering V_{th}, which wrongfully ignores the NM. Again, a slight difference in speed starts to develop as V_{th} decreases between the MD-DCVS and the single-V_{th} DDCVS with constant W_{keeper}. This is due to the increased loading at the DDCVS internal nodes (N_1 and N_2), because the keeper transistors are sized up to control the NM.

Figure 6.23 signifies that the MDDCVS consumes approximately the same dynamic power as the single V_{th} DDCVS with a controlled NM. This occurs because the SSG consumes very little dynamic power during the active mode, because no switching occurs.

The leakage power of the three cases is also shown in Figure 6.24. The MDDCVS has a substantially lower leakage consumption compared to that of the single V_{th} designs. This advantage is important to reduce the leakage currents of the HVT devices, and to enhance the speed by using LVT devices in the evaluation path. In Figures 6.22, 6.23 and 6.24, the HVT is also taken to be 550mV, whereas LVT is varied and denoted by V_{th}.

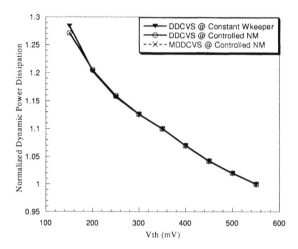

Figure 6.23.   Normalized dynamic power vs V_{th}

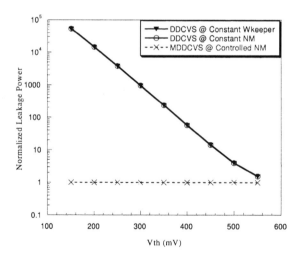

Figure 6.24.   Normalized leakage power vs V_{th}

6.7.   Chapter Summary

A modified Domino circuit, called the HS-Domino is developed. It resolves the trade-off between the performance and noise margins in conventional Domino logic. This circuit can now benefit from the scaling down of the technology and supply voltages, since it can tolerate the lower threshold voltages. The speed of the new Domino logic continues to improve as the threshold voltages are scaled down, while the noise margin is controlled. The

HS-Domino offers up to 60% energy savings compared to what a conventional domino offers. An MTCMOS implementation of the new Domino logic style is also devised. This dual-threshold implementation achieves low leakage values during standby, while high speed and low dynamic power are maintained during the active mode. The MTCMOS implementation of DDCVS logic gates also achieves substantially low leakage values during standby.

Notes

1 The CD-Domino is widely known in the industry as the D2 Domino, whereas conventional Domino is known as the D1 Domino.

2 In OR gates, only one input needs to be forced $high$, whereas other gates such as AND gates require all the inputs to be forced $high$.

References

[1] S. Thompson, I. Young, J. Greason, and M. Bohr, "Dual Threshold Voltage and Substrate Bias: Keys to High Performance, Low Power, 0.1μm Logic Designs," *in Digest of Technical Papers, IEEE Symposium on VLSI Technology*, pp. 69–70, June 1997.

[2] Z. Chen, C. Diaz, J. Plummer, M. Cao, and W. Greene, "0.18μm Dual Vt MOSFET Process and Energy-Delay Measurement," *in Proceedings of the International Electron Devices Meeting*, pp. 851–853, 1996.

[3] V. Oklobdzija and R. Montoye, "Design-Performance Trade-Offs in CMOS-Domino Logic," *IEEE Journal of Solid-State Circuits*, vol. SC-21, pp. 304–306, April 1986.

[4] M. Allam, M. Anis, and M. Elmasry, "High-Speed Dynamic Logic Styles for Scaled-Down CMOS and MTCMOS Technologies," *in Proceedings of the International Symposium on Low Power Electronics and Design*, pp. 155–160, July 2000.

[5] M. Anis, M. Allam, and M. Elmasry, "Energy-Efficient Noise-Tolerant Dynamic Styles for Scaled-Down CMOS and MTCMOS Technologies," *IEEE Transactions on VLSI Systems - Special Issue on Low Power Electronics and Design*, vol. 10, no. 2, pp. 71–78, April 2002.

[6] G. Yee and C. Sechen, "Clock-Delayed Domino for Dynamic Circuit Design," *IEEE Transactions on VLSI Systems*, vol. 8, no. 4, pp. 425–430, April 2000.

[7] Z. Chen, L. Wei, and K. Roy, "Estimation of Standby Leakage Power in CMOS Circuits Considering Accurate Modeling of Transistor Stacks," in *Proceedings of the International Symposium on Low-Power Electronics and Design*, August 1998, pp. 239–244.

[8] J. Kao and A. Chandrakasan, "Dual-Threshold Voltage Techniques for Low-Power Digital Circuits," *IEEE Journal of Solid-State Circuits*, , no. 7, pp. 1009–1018, July 2000.

[9] J. Kao, "Dual Threshold Voltage Domino Logic," *in Proceedings of the IEEE 25th European Solid-State Circuits Conference*, pp. 118–121, September 1999.

[10] P. Ng, P. T. Balsara, and D. Steiss, "Performance of CMOS Differential Circuits," *IEEE Journal of Solid-State Circuits*, vol. 31, no. 6, pp. 841–846, June 1996.

Further Reading

Leakage-Tolerant Dynamic Circuits

- A. Alvandpour, P. Larsson-Edefors, and C. Svensson, "A Leakage-Tolerant Multi-Phase Keeper for Wide Domino Circuits," *in Proc. IEEE International Conference on Electronics, Circuits and Systems*, pp. 209-212, 1999.

- L. Wang, R. Krishnamurthy, K. Soumyanath, and N. Shanbhag, "An Energy-Efficient Leakage-Tolerant Dynamic Circuit Technique," *in Proc. IEEE International ASIC/SOC Conference*, pp. 13-16, September 2000.

- M. Anders, R. Krishnamurthy, R. Spotten, and K. Soumyanath, "Robustness of sub-70nm Dynamic Circuits: Analytical Techniques and Scaling Trends," *in Symposium on VLSI Circuits Digest of Technical papers*, pp. 23-24, June 2001.

- R. Krishnamurthy, A. Alvandpour, G. Balamurugan, N. Shanbhag, K. Soumyanath, and S. Borkar, "A 0.13μm 6GHz 256x32b Leakage-tolerant Register File," *in Symposium on VLSI Circuits Digest of Technical papers*, pp. 25-26, June 2001.

- R. Krishnamurthy, A. Alvandpour, G. Balamurugan, N. Shanbhag, K. Soumyanath, and S. Borkar, "A 130-nm 6-GHz 256x32 bit Leakage-Tolerant Register File," *IEEE Journal of Solid-State Circuits*, VOL.37, NO.5, pp. 624-632, May 2002.

- A. Alvandpour, R. Krishnamurthy, K. Soumyanath, and S. Borkar, "A Sub-130nm Conditional Keeper Technique," *IEEE Journal of Solid-State Circuits*, VOL.37, NO.5, pp. 633-638, May 2002.

- S. Heo and K. Asanovic, "Leakage-Biased Domino Circuits for Dynamic Fine-Grain Leakage Reduction," *in Symposium on VLSI Circuits Digest of Technical papers*, pp. 316-319, June 2002.

- F. Hamzaoglu and M. Stan, "Circuit-Level Techniques to Control Gate Leakage for sub-100nm CMOS," *in Proc. IEEE International Symposium on Low Power Electronics and Design*, pp. 60-63, August 2002.

MTCMOS Dynamic Circuits

- H. Yoo, "Dual-V_T Self-Timed CMOS Logic for Low Subthreshold Current Multigigabit Synchronous DRAM," *IEEE Transactions on Circuits and Systems-II: Analog and Digital Signal Processing*, VOL.45, NO.9, pp. 1263-1271, September 1998.

- S. Jung, K. Kim, and S. Kang, "Transistor Sizing for Reliable Domino Logic Design in Dual Threshold Voltage Technologies," *in Proc. ACM Great Lakes Symposium on VLSI*, pp. 133-138, 2001.

- K. Fuji, T. Douseki, and Y. Kado, "A Sub-1V Dual-Threshold Domino Circuit Using Product-of-Sum Logic," *in Proc. IEEE International Symposium on Low Power Electronics and Design*, pp. 259-262, August 2001.

- W. Chen, W. Huang, P. Kudva, G. Gristede, S. Kosonocky, and R. Joshi, "Mixed Multi-threshold Differential Cascode Voltage Switch (MT-DCVS) Circuit Styles and Strategies for Low Power VLSI Design," *in Proc. IEEE International Symposium on Low Power Electronics and Design*, pp. 263-266, August 2001.

- S. Jung, K. Kim, and S. Kang, "Noise Constrained Transistor Sizing and Power Optimization for Dual V_t Domino Logic," *IEEE Transactions on VLSI Systems*, VOL.10, NO.5, pp. 532-541, October 2002.

- S. Jung, K. Kim, and S. Kang, "Low-Swing Clock Domino Logic Incorporating Dual Supply and Dual Threshold Voltages," *in Proc. ACM/IEEE Design Automation Conference*, pp. 467-472, June 2002.

- S. Jung, K. Kim, and S. Kang, "Timing Constraints for Domino Logic Gates with Timing-Dependent Keepers," *IEEE Transactions on Computer-Aided Design of Integrated Circuits and Systems*, VOL.22, NO.1, pp. 96-104, January 2003.

- K. Oh and L. Kim, "A Clock Delayed Sleep Mode Domino Logic for Wide Dynamic OR Gate," *in Proc. IEEE International Symposium on Low Power Electronics and Design*, August 2003. (To appear)

Chapter 7

MTCMOS CURRENT-STEERING CIRCUITS

7.1.   MOS Current Mode Logic: Overview

MOS Current Mode Logic (MCML) is well known as a high performance logic family for mixed signal systems [1] [2]. Figure 7.1 illustrates the architecture of an MCML inverter/buffer. Transistor Q_1 acts as a DC current source controlled by a bias voltage V_{bias}, and R are pullup resistors. The logic function is implemented by the logic block connected between the resistors and the current source. For an inverter/buffer, the logic block is the differential pair constructed by transistors Q_2 and $Q_2{}'$.

The operation of the MCML logic is based on the differential pair circuit; each input variable is connected to a differential pair circuit. The value of the input variable controls the flow of current through the two branches of the differential pair. For example, if $V_{gs}(Q_2)$ is higher than $V_{gs}(Q_2{}')$, the current flowing through Q_2 exceeds the current passing through $Q_2{}'$. Therefore, the voltage of node OUT begins to drop, until it reaches a steady state, where the current flowing through resistor R matches the current passing through transistor Q_2. At the same time, node $\overline{OUT}$ is charged to V_{dd}.

To achieve the best performance, all of the current must pass through the *on* branch only, and the load resistors should be small in order to reduce the RC delay. This guarantees that the voltage at one of the output nodes is V_{dd}, while the other nodal voltage is $V_{dd} - I_{bias}.R$, where I_{bias} is the value of the current flowing through the current source, and R is the load resistance. Thus, the output voltage swing of the MCML gate is $\Delta V = I_{bias}.R$. It is beneficial to have a small voltage swing because: (1) the gate delay is reduced, (2)

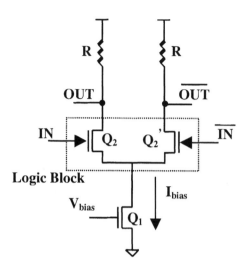

Figure 7.1.   Inverter/buffer implemented in MCML

crosstalk between adjacent signals is reduced, and (3) dynamic power dissipation is reduced in the case of long buses. Therefore, MCML can be used in the implementation of bus transceivers to reduce power [3] [4].

However, a very small ΔV is not recommended, because it renders the gate more susceptible to noise, and it reduces the current difference between the differential branches which means smaller discharge current. Typically, a swing of 20% of V_{dd} is used [2].

But, MCML has some major drawbacks. First, a large amount of static power is dissipated due to the constant current source. By using Power/MHz as a measure for the power dissipation, the MCML power dissipation is reasonable at high operating frequencies, but the Power/MHz becomes much higher at lower operating frequencies, because the current source is fixed. Compared to CMOS circuits, the MCML consumes more power at low frequencies. Therefore, the MCML is preferred in only high frequency applications, in order to reduce the overhead of the MCML's static biasing power. Secondly, MCML is not suitable for power-down modes because of the DC current source. Hence, it is inappropriate for large systems, where power down techniques are used to reduce the system power. Thirdly, MCML designs need to include a reference voltage distribution tree to control the current source of each gate, leading to larger chip area and more complex routing. Finally, the matching of the rise and fall delays is not an easy task, because it is a function of the load of each gate.

As a solution for some of the MCML problems in digital design, Mizuno et al. [1] have suggested an adaptive pipelined technique for MCML circuits. The new implementation changes the current source value of the gates in the critical path. The gates that are not in the critical path can use a smaller current source. This scheme reduces the overall power dissipation.

To summarize, the advantages of MCML are its:

- High speed because of the small output voltage swing;

- Low switching noise because of the constant current source;

- High noise immunity because of its differential nature;

- Efficient implementation of arithmetic circuits (XOR styles);

- Small output swing efficiency in driving data buses, and high load signals which reduces dynamic power dissipation and delay.

MCML's disadvantages are:

- DC current source wastes power during standby which is not suitable for low-power digital circuits;

- Mismatch between resistive loads;

- Needs a special tree for the reference voltage V_{bias} distribution needs to be added which increases the complexity of the layout stage.

It is clear that most of the problems associated with MCML are caused by the static current source and the matching of the loads.

7.2. Introduction

As CMOS process technology scales, the power consumption of future ICs will increase over time, if significant architectural changes are not made. Therefore, it is critical in future circuits that power be minimized beyond the traditional constraints of packaging costs and heat dissipation. A digital circuit style that seems to be promising in both reducing power consumption at high frequencies and providing high performance is MOS Current Mode Logic (MCML) [1] [4] [5] as mentioned in Section 7.1. MCML achieves high speeds due to the reduced output swings at its differential outputs. The output swings between V_{dd} and V_{dd}-ΔV, where V_{dd} and ΔV are the supply voltage and voltage output swing, respectively.

The two-level MCML circuit, in particular, is very popular. It consists of two series stacked NMOS devices (Q_2 and Q_3 as in Figure 7.2). For example, latches and D-flip-flops have been frequently used in the design of multiplexers, demultiplexers, and frequency dividers in high-speed optical communication circuits [6]. In addition, the basic 2-input logic gates employ a two-level MCML structure, for example, AND2, NAND2, OR2, and NOR2, which are shown in Figure 7.2, as well as XOR2 and XNOR2. Particularly, the XOR2 has been efficiently used in the implementation of high-speed microprocessor designs [1] [4].

As mentioned in Section 7.1, though, one of the drawbacks of MCML design is its high static power consumption. Power dissipation in MCML is equal to $V_{dd} \times I_{bias}$, where I_{bias} is the current flowing through the constant current source Q_1 (Figure 7.2). Power dissipation is attributed to several factors: (1) I_{bias} cannot be reduced under a certain limit; otherwise, performance will be degraded, (2) V_{dd} has a minimum value V_{min} to ensure that the differential pairs operate in the saturation region, and thus limit power reductions to the value of V_{min} (this will be further illustrated in Section 7.3), and (3) voltage level shifters are often employed in two-level MCML designs to reduce the input voltage signal (V_a in Figure 7.2) entering the lower level in order to guarantee that the differential pairs (Q_2 and $Q_2{'}$, Q_3 and $Q_3{'}$) operate in the saturation region. These level shifters contribute to a large power overhead for the MCML circuit, which will be illustrated in Section 7.4. However, the constant static current source achieves excellent noise immunity [4]. Given the three limitations, a method must be devised to reduce power consumption in MCML designs, while maintaining the constant current source to attain adequate noise immunity.

In this chapter, a Multi-Threshold CMOS (MTCMOS) implementation of MCML is introduced. By carefully assigning threshold voltage values to the differential pairs for both levels of the MCML, a further reduction in the operating supply voltage is possible. Furthermore, the MTCMOS implementation ensures that the differential pair devices, previously operating in the linear mode, can operate now in the saturation region without the need of level shifters. Then, power dissipation can be significantly reduced. Over the past decade, MTCMOS technology has been used to reduce leakage during the standby mode, while attaining high speed in the active mode [7] [8].

However, the uniqueness of this work is that the MTCMOS implementation will not be used to reduce leakage power in the standby mode. Instead, the MTCMOS implementation will be used to reduce the power supply voltage, and ensure the correct functionality in the active mode. To illustrate the power efficiency of the MTCMOS implementation, this chapter will first explain the reasons why high power is dissipated in conventional MCML designs. The chapter will begin with the calculation of the minimum supply voltage V_{min} for a conventional two-level MCML design, followed by an explanation of the need for level shifters.

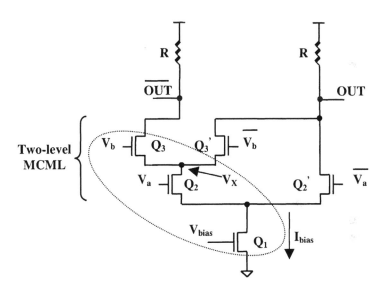

Figure 7.2. AND2/NAND2/OR2/NOR2 MCML gate

7.3. Minimum Supply Voltage: First Constraint

The minimum operating supply voltage V_{min} for an MCML circuit is defined as the lowest voltage at which the differential pairs at both levels (Q_2 and Q_2', Q_3 and Q_3'), as well as the current source Q_1, are operating in the saturation region to ensure correct functionality and adequate performance. From Figure 7.2, V_{min} can be expressed as

$$V_{min} = V_{ds_{1_{SAT}}} + V_{ds_{2_{SAT}}} + V_{gs_3} \qquad (7.1)$$

where $V_{ds_{1_{SAT}}}$, $V_{ds_{2_{SAT}}}$ and V_{gs_3} are the drain-source-saturation voltage for Q_1, the drain-source-saturation voltage for Q_2, and the gate-source voltage for Q_3, respectively. In general, the drain-source-saturation voltage $V_{ds_{SAT}}$ is expressed as

$$V_{ds_{SAT}} = V_{gs} - V_{th} \tag{7.2}$$

where V_{th} is the threshold voltage. In the saturation region, V_{gs} can be expressed as

$$V_{gs} = \sqrt{\frac{I}{K}} + V_{th} \tag{7.3}$$

where I is the current flowing in the device, and K is a parameter function in the process transconductance and the width to length ratio of the device. Substituting in Equation (7.2), $V_{ds_{SAT}}$ can now be written as

$$V_{ds_{SAT}} = \sqrt{\frac{I}{K}} \tag{7.4}$$

From Equation (7.1), and if $I_{bias}=I_2=I_3$ is taken for the ideally conducting case (all the bias current flowing in the conducting branch) and $K_2=K_3=K$ for the two levels of differential pairs, then

$$
\begin{aligned}
V_{min} &= \sqrt{\frac{I_{bias}}{K_1}} + \sqrt{\frac{I_2}{K_2}} + \sqrt{\frac{I_3}{K_3}} + V_{th_3} \\
&= \sqrt{I_{bias}}\{\frac{1}{\sqrt{K_1}} + \frac{2}{\sqrt{K}}\} + V_{th_3} \\
&= \sqrt{K_1(V_{bias} - Vth_1)^2}\{\frac{1}{\sqrt{K_1}} + \frac{2}{\sqrt{K}}\} + V_{th_3} \\
&= (V_{bias} - Vth_1)\{1 + \frac{2\sqrt{K_1}}{\sqrt{K}}\} + V_{th_3} \\
&\approx 3V_{bias} - 3V_{th_1} + V_{th_3}
\end{aligned}
\tag{7.5}
$$

where V_{th_1} and V_{th_3} are the threshold voltages of Q_1 and Q_3, respectively. It is clear from Equation (7.5) that V_{min} can be reduced by attaining V_{th_1} at a high value and reducing V_{th_3}. This conclusion will be used at the end of Section 7.4 to justify the use of the proposed MTCMOS design.

7.4. Saturation Assurance: Second Constraint and the Proposed MTCMOS Design

In this section, the saturation conditions for the differential pair devices Q_2 and Q_2', Q_3 and Q_3' are first investigated. This will eventually explain the need for voltage level shifters in MCML designs. From Figure 7.2, Q_3 operates in the saturation region if

$$V_{gd_3} \leq V_{th_3} \tag{7.6}$$

and

$$V_b - (V_{dd} - \Delta V) \leq V_{th_3} \tag{7.7}$$

where V_b is the voltage input to Q_3. Equation (7.7) can be rewritten as

$$V_b \leq V_{dd} - \Delta V + V_{th_3}. \tag{7.8}$$

Therefore,

$$
\begin{aligned}
V_{b_{max}} &= min[V_{dd}, V_{dd} - \Delta V + Vth_3] \\
&= V_{dd}
\end{aligned}
\tag{7.9}
$$

for ΔV=400mV and V_{th_3}=500mV in 0.18μm CMOS technology. Thus, the condition of saturation is fulfilled, when V_b toggles from V_{dd} to V_{dd}-ΔV. Similarly, Q_2 operates in the saturation when

$$V_{gd_2} \leq V_{th_2} \tag{7.10}$$

giving

$$V_a - V_X \leq V_{th_2} \tag{7.11}$$

which is rewritten as

$$V_a \leq V_X + V_{th_2} \tag{7.12}$$

where V_X is the voltage of the intermediate point shown in Figure 7.2.

Now, taking the current I_{bias}=I_3 during conduction (if it is assumed that all I_{bias} flows in the conducting branch), then

$$K_1(V_{bias} - V_{th_1})^2 = K_3(V_b - V_X - V_{th_3})^2 \tag{7.13}$$

Therefore,

$$V_X = V_b - V_{th_3} - \sqrt{\frac{K_1}{K_3}}(V_{bias} - V_{th_1}) \tag{7.14}$$

Equation (7.12) can now be written as

$$V_a \leq V_b - V_{th_3} - \sqrt{\frac{K_1}{K_3}}(V_{bias} - V_{th_1}) + V_{th_2} \qquad (7.15)$$

Therefore,

$$V_{a_{max}} \approx V_b - V_{bias} + V_{th_1} - V_{th_3} + V_{th_2} \qquad (7.16)$$

for $K_1 \approx K_3$. For 0.18μm CMOS technology, V_{bias} is taken as 0.8V, and $V_{th_1} = V_{th_2} = V_{th_3} = 0.5$V, leading to

$$V_a \approx V_b - 0.3V \qquad (7.17)$$

Thus, level shifters are used to shift-down the voltage of the input V_a entering the lower level (Q_2) by ≈ 0.3V below V_b. V_b and $\overline{V_b}$ will, therefore, toggle from 1.8 to 1.4V (assuming $V_{dd} = 1.8$V and ΔV=400mV), while V_a and $\overline{V_a}$ will toggle from 1.5 to 1.1V (0.3V less than V_b) as denoted in Figure 7.3. This ensures that the differential pairs in both levels are operating in the saturation region, and consequently ensures correct operation. These level shifters dissipate a large amount of power. Therefore, a technique should be developed to (1) eliminate the usage of level shifters, (2) ensure that the differential pairs still operate in the saturation region, and (3) reduce the operating supply voltage V_{dd}. The use of MTCMOS technology to implement MCML solves these problems.

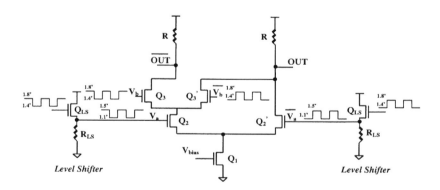

Figure 7.3.   MCML AND gate with level shifters

From Equation (7.16), by utilizing the MTCMOS technology, V_a is made higher by keeping V_{th_1} and V_{th_2} *high* while reducing V_{th_3}. This occurs because if V_{th_1} and V_{th_2} are kept *high* =0.5V, while V_{th_3} is made =0.2V, with

V_{bias}=0.8V, then V_a=V_b. Therefore, both V_a and V_b will toggle at the same voltage levels, while ensuring that all the differential pair devices operate in the saturation zone. Thus, the need to employ level shifters is no longer necessary, and the power dissipated by the level shifters can now be saved. Furthermore, the supply voltage has been further reduced, adding to the power savings. Referring to Equation (7.5), with V_{bias}=0.8V, V_{th_1}=V_{th_3}=0.5V, then V_{min}=1.4V. For the MTCMOS implementation, if V_{th_1} and V_{th_2} are kept *high*, while V_{th_3}=0.2V, this would result that V_{min}=1.1V. Therefore, a power saving of $\approx$ 22% is achieved. The schematic for the power-efficient MTCMOS MCML AND gate is drawn in Figure 7.4. The low-V_{th} (LVT) devices (Q_3 and $Q_3{}'$) are highlighted; the high V_{th} (HVT) (Q_2 and $Q_2{}'$, Q_1) are not.

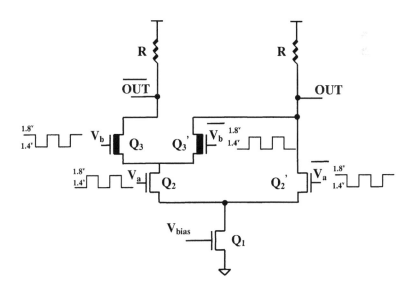

Figure 7.4. MTCMOS MCML AND gate

7.5. A 2.5 Gbit/s 1:8 Demultiplexer in MTCMOS MCML

The MTCMOS MCML implementation is applied to a 2.5 Gbit/s 1:8 Demultiplexer (DEMUX) to verify functionality. Figure 7.5 shows a block diagram of the 1:8 DEMUX.

It is a tree-type circuit made up of 1:2 DEMUX circuits. The 1:2 DEMUX in Figure 7.6 is composed of two D-Flipflops (DFFs): a Master-Slave (MS) and a Phase-Shifting (PS) DFF . The PS-DFF employs a subsequent latch in

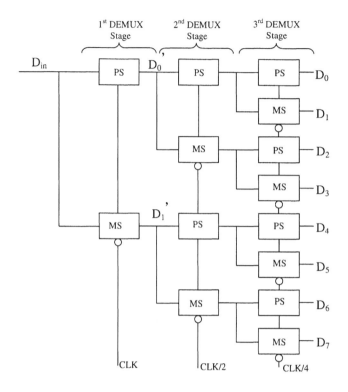

Figure 7.5.   1:8 Demultiplexer

its path to ensures that both outputs (D_0' and D_1') have transitions at the same time. The alternate operation of the two DFFs generates two data streams from the input data stream D_{in}. The 1:2 demultiplexer uses both edges of its input clock (CLK). The MS-DFF outputs odd data streams at the rising edges of the clock, while the PS-DFF outputs even data streams at the falling edges of the clock. These latched data are outputs at the rising edges of the clock by the second Master latch in the PS-DFF. This can be seen in the timing diagram of Figure 7.8. In this way, 2-bit output data (D_0' and D_1') are synchronized with the rising edges of the input clock. Because both edges of the clock are used, the input data rate is twice the input clock rate. The clock will, therefore, have a frequency of 1.25GHz. A clock frequency of CLK/2 and CLK/4, which are produced by a clock divider, are applied to the second and third DEMUX stages, respectively. Thus, the final data stream at the 1:8 DEMUX output has a data rate of 312.5 Mbit/s. The MTCMOS 1:2 D-latch is shown in Figure 7.7.

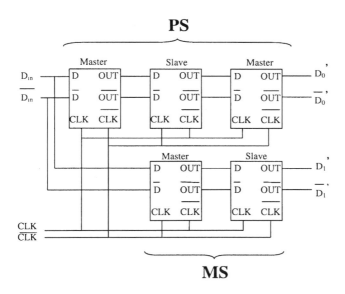

Figure 7.6. 1:2 Demultiplexer

The LVT is set to 200mV, whereas the HVT is set to 500mV.

In order to achieve a sufficient phase margin, the timing of the CLK, CLK/2 and CLK/4 signals should be set at the center of each D_0/D_1 eye as shown in Figure 7.8. Also, an MS and PS delay is shown which is an internal delay for the data to be evaluated at the 1:2 DEMUX output. By using a 0.18μm CMOS technology, the 1:8 DEMUX circuit is implemented in both conventional MCML and the proposed MTCMOS MCML. Both implementations are operated at their minimum voltage supplies, that is, 1.4V for a conventional MCML and 1.1V for MTCMOS MCML. In addition, the conventional MCML employs level shifters to ensure correct functionality and operation in the saturation region. As discussed in Section 7.4, level shifters are not needed in the MTCMOS design. While both implementations achieve the necessary throughput, the proposed MTCMOS design achieves a 37% reduction in power compared to the conventional design. Approximately 65% of the power savings is attributed to the reduction in the voltage supply, whereas the remaining 35% is attributed to the elimination of the level shifters.

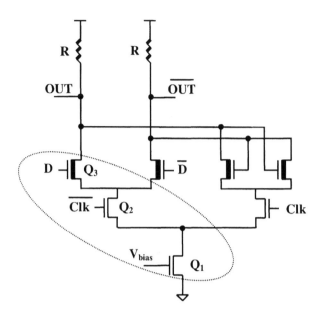

Figure 7.7.   D-Latch in MTCMOS MCML

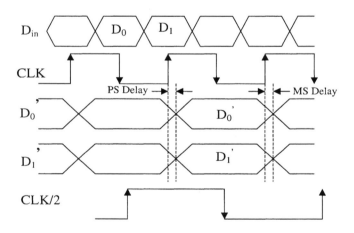

Figure 7.8.   Timing diagram

7.6.   Impact of Using MTCMOS Technology Over MCML Parameters

In the previous section, the use of MTCMOS to implement MCML designs has proved to be power efficient by allowing further reduction to the supply voltage, and the elimination of level shifters. However, the impact that the MTCMOS design may have on other circuit parameters such as gain, output resistance of the current source, and the common-mode-rejection-ratio should also be investigated. This is very important since MCML designs are parameter sensitive as most analog circuits are [9]. The following subsections will explore the impact that the MTCMOS design has on the circuit parameters; that is, what other effects are associated by keeping V_{th_1} and V_{th_2} high, and V_{th_3} low.

Delay, Power, PDP, and EDP

Delay: The delay of an MCML gate is expressed as $C_L \Delta V / I_{bias}$, where $I_{bias} = K_1(V_{bias} - V_{th_1})^2$, and C_L is the total load capacitance at the output. By keeping V_{th_1} high, $(W/L)_1$ should slightly increase to produce sufficient current, where $(W/L)_1$ is the width to length ratio of Q_1. A slight increase in the area of the single bias device, compared to the whole gate, will not augment the total area. On the other hand, it should be noted that increasing the current will come at the expense of power dissipation and poor Voltage Swing Ratio (VSR). VSR is the ratio between the output and input voltage swings . In an ideal MCML gate, all the bias current flows in the *on* branch (VSR=1); in reality, some finite amount of current flows in the *off* branch. Consequently, not all the bias current flows in the *on* branch. This nonideality causes a reduction in the output voltage swing (VSR<1). In conclusion, the MTCMOS implementation does not effect the speed of the circuit.

Power: With the additional reduction in the required supply voltage value due to the MTCMOS implementation, power dissipation can be reduced. A power saving of $\approx 22\%$ is achieved. Furthermore, power is saved due to the elimination of voltage level shifters. As a result of the reduction in power dissipation, both the Power-Delay-Product (PDP) and Energy-Delay-Product (EDP) are reduced [1].

It should also be noted that the LVT devices (Q_3 and $Q_3{}'$) increase the leakage current. This is not a problem, because the leakage current of MCML circuits can be included in the constant static current source.

Output Impedance

The output impedance r_o of an MOS device is expressed as

$$r_o = \frac{1}{\lambda I} = \frac{1}{\lambda K (V_{gs} - V_t)^2} \tag{7.18}$$

where λ is the output impedance constant. Therefore, the output impedance of the current source $r_{o_1} = \frac{1}{\lambda_1 K_1 (V_{bias} - V_{th_1})^2}$. r_{o_1} is shown in Figure 7.11. Then, it can be deduced that if V_{th_1} remains high, so will r_{o_1}. A large output impedance r_{o_1} is appealing for three main reasons:

- **High Current Matching Ratio**

  With a high output impedance, the actual current flowing through the current source is very close to the reference current source. From Figure 7.9,

$$I_{ref} = K_{ref}(V_{gs_{ref}} - V_{th_{ref}})^2(1 + \lambda_{ref}V_{ds_{ref}}) \qquad (7.19)$$

  and

$$I_{bias} = K_1(V_{gs_1} - V_{th_1})^2(1 + \lambda_1 V_{ds_1}) \qquad (7.20)$$

  $V_{gs_{ref}}$ is equal to V_{gs_1}, and if Q_{ref} and Q_1 are matched, then $K_{ref}=K_1$, $V_{th_{ref}}=V_{th_1}$, and $\lambda_{ref} = \lambda_1 = \lambda$. Therefore, from Equation (7.19) and Equation (7.20), the following can be written:

$$\frac{I_{bias}}{I_{ref}} = \frac{1 + \lambda V_{ds_1}}{1 + \lambda V_{ds_{ref}}} \qquad (7.21)$$

  Since $r_o \propto \frac{1}{\lambda}$ (Equation (7.18)), as r_o remains high, λ will remain low, resulting in $I_{bias} \approx I_{ref}$, and achieving a high current matching ratio.

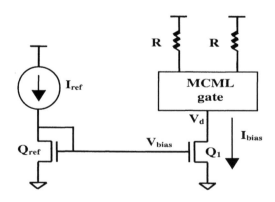

Figure 7.9.   Current Matching

- **Pure Current Source**

With r_{o_1} high, the drain voltage for the current source (V_d in Figure 7.9), would be stabilized (virtually constant) when Q_1 operates in the saturation zone. If r_{o_1} is reduced, this would cause V_d to fluctuate, which may cause Q_1 to conduct in the triode region. This effects the functionality of MCML, and ultimately degrades its speed. Therefore, a large r_{o_1} is desired.

- **Common Mode Rejection Ratio**

The Common Mode Rejection Ratio (CMRR) defines the ability of an MCML gate to reject common-mode signals (noise) at its inputs so as not to disrupt the output. The higher the CMRR, the more immune it is to common-mode noise. CMRR is proportional to r_{o1} [9]. Therefore, if r_{o1} remains high, so will the CMRR. Very high values for the CMRR ensure immunity against common mode noise, reducing the chances for the MCML to malfunction. A rise in the CMRR will also allow the utilization of smaller voltage swings which then enhances speed. Furthermore, a large CMRR will compensate for any effect the V_{th} fluctuation (mismatch) of the differential pairs may have on the output amplitude of the MCML circuit [6].

Gain

In the MTCMOS implementation of MCML, illustrated in Section 7.4, V_{th_3} is reduced by taking a value of 200mV, causing r_{o_3} to decrease (Equation (7.18)), whereas r_{o_1} and r_{o_2} are $high$, because V_{th_1}, V_{th_2} remain $high$. Therefore, the DC differential gain $G(0)$ of the MCML circuit can be approximated as

$$G(0) = G(0)_{ideal} \frac{r_{o_3} R}{r_{o_3} + R} \qquad (7.22)$$

where $G(0)_{ideal}$ is the ideal DC differential voltage gain ($r_{o_1} = r_{o_2} = r_{o_3} = \infty$), and R is the load resistance. r_{o_3} in Equation 7.22 is the output impedance of the LVT differential pair (Q_3 and Q_3'). In the MTCMOS design, V_{th_3} is reduced from 500 to 200mV, causing r_{o_3} to reduce by 34% which subsequently should reduce the circuit gain $G(0)$ (Equation (7.22)). However, given that r_{o_3} is in the order of 300KΩ, while R is 300Ω, the gain reduction is $\approx 0.0006\%$ which is negligible.

It should be mentioned that a high gain strengthens the regeneration and bistability of the MCML gate. However, MCML does not a require high gain such as that of the static CMOS case, because most of the noise which adversely affects CMOS circuits becomes common mode noise for the MCML, and is rejected by the differential logic. Since MCML circuits also generate significantly less switching noise than CMOS circuits, then the environment will be more conductive to a low gain operation. A low gain is also appealing so as not to amplify the undesired effect of any process variation, or mismatch associated with the differential pairs or resistive loads [6] as will be shown next. A gain just over "1" is sufficient to achieve regeneration, bi-stability, and immunity against process, voltage and mismatching conditions.

V_{th} Fluctuations

A V_{th} mismatch in the differential pair transistors causes the output waveform to be unbalanced. An offset bias voltage ΔV_B at the output exists, which is given by [6]

$$\Delta V_B = G(0)\Delta V_{th} \qquad (7.23)$$

where ΔV_{th} is the fluctuation in V_{th}. From Equation (7.23), $\Delta V_B \propto G(0)$. Therefore, a small DC gain is preferable to reduce the effect of the mismatch. It was shown that the DC differential gain $G(0)$ is not varied, and therefore, has no effect over ΔV_B compared to conventional MCML, leading to a similar response for the V_{th} fluctuations.

Frequency Response

Since MCML is used in high-speed applications such as optical communication circuits, analyzing its frequency response is important. Figure 7.11 illustrates the impact high frequency has on the differential-mode and common-mode gains.

- Differential-mode gain

    Because the MCML circuit is a simple source-coupled pair circuit, differential-mode voltage gain $G(f)$ at a frequency f is generally given by [10]

$$G(f) = \frac{G(0)}{1 - j(2\pi f/p)} \qquad (7.24)$$

    where $G(0)$ is the DC gain of the circuit, and p is the pole of the circuit. The pole is a function of the load resistance R and load capacitance C_L of Q_2 or Q_2' (Figure 7.11). From Equation (7.24), as the frequency increases, the differential-mode gain decreases. The simulated differential-mode gain as a function of frequency is portrayed in Figure 7.10. Since the pole p is not a function of V_{th_1}, V_{th_2} or V_{th_3}, then $G(f)$ is impacted in the same way as $G(0)$, which was discussed previously.

- Common-mode gain

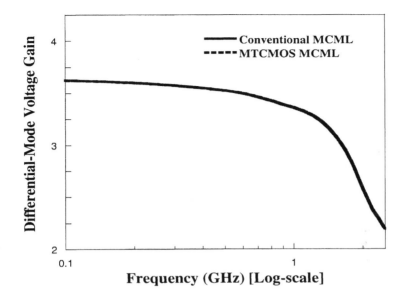

Figure 7.10. Frequency response of D-latch in conventional and MTCMOS MCML

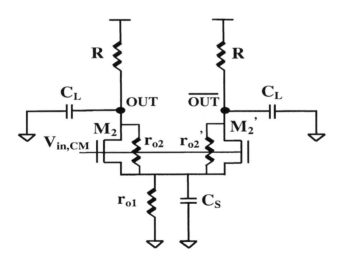

Figure 7.11. Equivalent common-mode circuit at high frequency

The common-mode gain G_{CM} at a mismatch between Q_2 and Q_2' is expressed as [9]

$$G_{CM} = \frac{\Delta gm_2 [R \parallel (\frac{1}{C_L s})]}{(gm_2 + gm_2')[r_{o_1} \parallel (\frac{1}{C_s s})] + 1} \tag{7.25}$$

where C_s is the capacitance contributed by Q_1, gm_2 and gm_2' are the transconductance of Q_2 and Q_2' respectively, and Δgm_2 is the mismatch difference between gm_2 and gm_2'. Equation (7.25) suggests that G_{CM} will be low since r_{o_1} remains high. Low G_{CM} values are desirable to increase the CMRR, and ultimately to be immune to common-mode noise.

7.7. Chapter Summary

This chapter presented an MTCMOS implementation of MOS Current Mode Logic. By using a 1:8 2.5Gbit/s demultiplexer as a test vehicle, a 37% power saving is achieved while the required data rate is attained. Furthermore, the MTCMOS design shows no impact on circuit parameters such as output impedance, CMRR, gain, V_{th} fluctuations, and frequency response.

References

[1] M. Mizuno, M. Yamashina, K. Furuta, H. Igura, H. Abiko, K. Okabe, A. Ono, and H. Yamada, "A GHz MOS Adaptive Pipeline Technique Using MOS Current-Mode Logic," *IEEE Journal of Solid-State Circuits*, vol. 31, no. 6, pp. 784–791, June 1996.

[2] M. Yamashina and H. Yamada, "MOS Current Mode Logic (MCML) Circuit for Low-Power GHz Processors," *NEC Research & Development*, vol. 36, no. 1, pp. 54–63, January 1995.

[3] P. Ng, P. T. Balsara, and D. Steiss, "Performance of CMOS Differential Circuits," *IEEE Journal of Solid-State Circuits*, vol. 31, no. 6, pp. 841–846, June 1996.

[4] M. Yamashina and H. Yamada, "An MOS current mode logic (MCML) circuit for Low-Power Sub-GHz Processors ," *IEICE Trans. Electron*, vol. E75-C, pp. 1181–1187, October 1992.

[5] J. Musicer and J. Rabaey, "MOS Current Mode Logic for Low Power, Low Noise CORDIC Computations in Mixed-Signal Environments," in *in Proceedings of the International Symposium on Low-Power Electronics and Design*, 2000, pp. 102–107.

[6] A. Tanabe, M. Umetani, I. Fujiwara, T. Ogura, K. Kataoka, M. Okihara, H. Sakuraba, T. Endoh, and F. Masuoka, "0.18-μm CMOS 10-Gb/s Multiplexer/Demultiplexer ICs Using Current Mode Logic with Tolerance to Threshold Voltage Fluctuation," *IEEE Journal of Solid-State Circuits*, vol. 36, no. 6, pp. 988–996, 2001.

[7] L. Wei, Z. Chen, K. Roy, M. Johnson, and V. De, "Design and Optimization of Dual-Threshold Circuits for Low-Voltage Low-Power Applications," *IEEE Transactions on VLSI Systems*, vol. 7, no. 1, pp. 16–24, March 1999.

[8] S. Mutah, T. Douseki, Y. Matsuya, T. Aoki, S. Shigematsu, and J. Yamada, "1-V Power Supply High-Speed Digital Circuit Technology with Multi-Threshold Voltage CMOS," *IEEE Journal of Solid-State Circuits*, vol. 30, no. 8, pp. 847–853, August 1995.

[9] B. Razavi, *Design of Analog CMOS Integrated Circuits*, McGraw-Hill Ryerson Ltd, 2000.

[10] P. Gray and R. Meyer, *Analysis and Design of Analog Integrated Circuits*, New York: Wiley, 1993.

Index